Corporate Commu[nications]
in Construct[ion]

Public Relations Strategies
for Successful
Business and Projects

CHRISTOPHER N. PREECE
BSc(Hons), PhD, MCIOB
Construction Management Group, School of Civil Engineering, University of Leeds

KRISEN MOODLEY
BSc(Hons), MSc
Construction Management Group, School of Civil Engineering, University of Leeds

ALAN M. SMITH
BA(Hons)
Head of Public Relations, Edmund Nuttall Limited

b

**Blackwell
Science**

Contents

Preface   v
Acknowledgements   vii
Introduction   ix

Part One   **1**

1   Corporate communications and stakeholders   3
2   Corporate identity – not just pretty colours and logos!   16
3   Corporate communications management: objectives, skills
    and organisation   33

Part Two   **41**

4   Marketing communications – caring for the client   43
5   Media relations – what the papers say   60
6   Financial relations – money talks   72
7   Government relations – 'the powers that be'   79
8   Communicating community involvement –
    what will the neighbours say?   100

Part Three   **117**

9   Internal communications – keeping our own house in order   119

Part Four   **137**

10 Corporate communications and the role of the project
    manager   139
11 Communicating safety in construction   152
12 Crisis management – 'That's another fine mess…!'   192

Part Five   **203**

13 Conclusions   205

References	213
Bibliography	218
Index	221

Preface

The construction industry needs to communicate much better with the outside world. A number of high profile reports to the industry, most notably the work of Sir Michael Latham, have highlighted the need for construction to improve its image with a wide variety of audiences, particularly clients, the media and communities. Initiatives have been frustrated by the highly fragmented nature of the industry. A lack of a clear, co-ordinated and coherent message has meant that a vitally important industry is largely misunderstood and viewed in a negative light by many people.

All organisations in construction need to focus on managing communications and public relations. This book identifies the key processes, tools and techniques that companies and their managers can use at the corporate and project level of their business. These include:

- Implementation of corporate identity as a catalyst for positive change.
- Development of more effective marketing communications and client care.
- Approaches to improve media relations.
- Building of better relationships with the financial community.
- Making the case for construction with government.
- Development of community policies which can improve the image of the construction business, win work and help develop the skills of the construction team.
- Development of communications skills of the project management team.
- Practical communication of safety policies and programmes which help avoid accidents and thus damaging publicity which can affect marketing success.

The principles highlighted in this text have been well established in industry and commerce generally. Their application in the highly competitive environment of construction has never been more crucial.

Christopher N. Preece
Krisen Moodley
Alan M. Smith

Acknowledgements

The authors would like to pay tribute to their respective families, friends and colleagues who have supported them during the course of writing this book.

Thanks should also be extended to many colleagues within organisations in the construction industry who have provided invaluable advice and information to support this work.

Alan Smith would especially like to thank Deborah Emberson whose patience and assistance in producing his contribution to this book have been invaluable.

Introduction

What is corporate communications?

The phrase 'corporate communications' means different things to different people. A design company may see it as meaning visual media; a form of corporate identity expressed through such things as logos or vehicle livery. A public relations consultancy or advertising agency may interpret it as aspects of publicity in editorial or paid space in magazines and newspapers. But increasingly the two words have come to mean what an organisation signals about itself: how it presents itself to its customers, suppliers, shareholders, competitors, staff, prospective recruits and the world at large.

It is obviously desirable that the various perceptions that people have about a company's 'image' should be favourable. The image should be broadly consistent, but, more importantly, it should reflect reality and the corporate culture of the organisation. For this to be achieved, an organisation must take a proactive approach to managing the way it 'communicates'.

Managing relationships more effectively

Corporate communications is fundamentally about managing relationships more effectively. Any business organisation comes into contact with a wide variety of different people and organisations with whom it needs to relate positively for its own commercial success.

This vital management function needs to be achieved within the context of the overall strategic management of the corporation. It needs to be planned and co-ordinated in line with the broader mission, objectives and strategies of the organisation. It may require the setting up of a department in larger organisations, or the allocation of specific duties to a senior manager in smaller companies. Commonly, it is the public relations department and managers who are charged with these responsibilities, but, increasingly, all managers throughout the business need to be better communicators.

In addition to corporate communications being afforded appropriate levels of resources in terms of time, skilled personnel and money, it needs to become part of the culture of the business. The way people communicate and relate to each other within the boundaries of the organisation, and with the different external groups within the public, is crucial to business success.

This text aims to examine how corporate communications processes and techniques may be managed more effectively in construction. Whilst the basic principles of corporate communications are the same for any industry, the book will look at how these may be adapted to create effective strategies within the dynamic and often turbulent environment of construction.

What are the distinct challenges to the construction business?

In many respects the very nature of the construction industry places distinct pressures on those charged with managing public relations and ensuring effective corporate and marketing communications. These challenges are concerned largely with how the organisation communicates and relates effectively with its many and diverse audiences. The challenges may be summarised as follows:

- Construction is largely project-based and highly mobile. Its impact on the economic, social and environmental life of society is considerable. The extension to Manchester Airport and Newbury Bypass are illustrations of this impact and the effects it has on a local, national and even international scale.
- Project organisations are often highly complex, involving the management and co-ordination of many different organisations involved in the construction process.
- At the level of the construction project, management is about managing relatively temporary teams of designers, engineers, builders and manufacturers for short periods of time.
- The nature of construction work on site inevitably involves people working in dangerous and hazardous circumstances. The safety and health of the people on site, and indeed of those living and working in the immediate vicinity, are of paramount concern.
- The industry is historically and inherently a conflict-orientated pro-

fessional culture. The division of people along the lines of their professional training has caused a fragmentation which has consequences for the way in which the industry is perceived by those who are not party to its internal politics.

A dynamic business environment

The construction industry over the last few decades has seen remarkable changes which have forced business organisations to consider their overall management approach. A number of these changes in the business environment have consequences for corporate communications. These may be stated as follows:

- There is a need to ensure the satisfaction of increasingly demanding and knowledgeable customers and clients.
- Partnering, joint venture arrangements and strategic alliances are all about how people and the organisations they represent relate to and communicate with one another.
- Relationship marketing demands improved communications methods and techniques between the customer, their professional advisors and the construction organisation.
- The Private Finance Initiative (PFI) and new contractual arrangements such as Design, Build, Finance and Operate (DBFO) provide for a fuller set of responsibilities and a wider role, particularly for the contractor.
- The introduction of new environmental legislation and greater awareness by clients create a need to communicate a more responsible approach.
- Increased general public awareness of environmental issues and direct action by pressure groups require a more proactive strategy for handling events and preventing potentially very damaging publicity.
- There is a need to improve relations with critical broadcast and print media.
- The need to retain and maintain numbers of skilled people in the industry means that there has to be a better understanding of their needs and motivations by employers.
- There has been a drain of skills owing to the loss of many skilled and professional people during the recent recession.
- A positive approach to relations with local communities in which the organisation operates will increase the efficiency of projects and improve the quality of life for all parties concerned.
- Government continues to be a major influence on the construction

industry despite a shrinkage of public sector demand. Improved relations with policy makers at local, regional, national, European and international levels are therefore essential.

- As construction companies have become larger due to acquisitions and mergers, both domestically and internationally, or have begun to seek business opportunities on a 'global' level, there is a greater need for co-ordinated and planned strategies for communications.

These potential communication challenges will be addressed in further detail throughout this text.

Taking a planned approach

Given the challenges outlined, corporate communications needs to be approached as seriously as any other management function. It needs to be seen as an integral part of both corporate and project-level decision making. Managers and staff at all levels, and within all functions, need to be aware of the communications consequences of what they say and what they do.

This text will examine what construction businesses need to achieve in terms of improved corporate communications with their various audiences; the tools and techniques that are most appropriate; and the management organisation and skills necessary for effective implementation.

What's in it for us?

A planned and proactive approach to the management of communications will provide a number of tangible benefits for the construction organisation. It should:

- gain the support of management and staff for necessary internal organisational changes and marketing efforts;
- ensure a more positive business environment and enable clients and their advisors to make more informed decisions when selecting companies for projects;
- improve relations with all parties involved in projects so that they can deliver a high quality of service to the total satisfaction of the client;
- attract investment and confidence from the City;
- gain the support and understanding of government as the regulator of construction legislation that affects demand for projects, and as a major potential client;

- assist the media in their search for information on which to base their stories and hopefully ensure more positive perceptions.

Structure of the book

This text is divided into five parts which aim to provide an insight into the management of public relations strategies at corporate and project levels in construction.

Part One introduces and defines the functions of corporate communications management.

Chapter 1 commences by *identifying the stakeholders* of the construction business. These are the people and organisations to whom the business needs to relate positively, and who are the targets for communications strategies. Being able to identify the primary and secondary stakeholders and their values, interests and activities and how these may impact the organisations will enable a more proactive approach to communications. This will ensure that the most appropriate media and messages are designed and used.

Chapter 2 examines the importance of *developing and managing a corporate identity and image* for the business. This will be shown to be much more than a cosmetic exercise, but a vital strategic management tool. Given the highly dynamic nature of construction, this is particularly challenging. Restructuring of organisations, acquisitions, mergers and take-overs, together with globalisation of construction companies, place even greater pressure on the management of organisations to ensure identification and loyalty of its own employees, customers and others.

Chapter 3 goes on to identify *how companies should organise themselves* to ensure better management of the public relations and corporate communications function. It discusses the objectives, skills and management decisions that may be shared by all people in the organisation, from the board of directors to the site management team. It is stressed that although public relations and corporate communications are 'everybody's business', it is necessary in most organisations to have a separate department dedicated to the planned management of these processes. Whether this should be part of existing arrangements for marketing or a distinct function is debated. Consideration is given to the possible need to employ specialist help from consultants and agents for specific tasks, for example media, financial or government relations, or for techniques which require expert assistance such as the design of brochures or advertisements, corporate hospitality and sponsorship. Having executed strategies which are aimed at improving perceptions and building relationships, some form of evaluation of their effectiveness is necessary.

Part Two examines in more detail the corporate communications role in communicating with a range of external audiences.

Chapter 4 focuses on the most important stakeholders of the construction business – the clients. The chapter emphasises the need to communicate a caring approach, and to deliver a high quality service. Techniques include the building of relationships and partnerships and the provision of well targeted and tailored promotional literature which addresses clients' needs and the demands of any proposed project as part of pre-qualification documentation.

Chapter 5 examines *the need for an effective approach to the media* at different levels. Building relationships with editors and journalists in the print and broadcast media may ensure fairer and more accurate treatment. But nothing should be taken for granted. A strategy which builds a professional working relationship is preferable to mistrust.

Chapter 6 looks at the highly specialised area of financial communications with shareholders, investment analysts, financial editors and regulators. The financing of any commercial organisation is of considerable importance and construction is no exception. Clients are concerned to check the standing of prospective consultants/contractors whom they may employ on projects. The chapter identifies the various financial stakeholders and their communications needs. Emphasis is placed on the careful planning and design of the annual report and accounts as a principal communications document.

Chapter 7 focuses on *the need for companies to communicate with government* as legislators, funding bodies and potential clients. Emphasis is on approaching Parliament in Westminster, but many of the techniques will apply at other levels: for example, finding out who makes and influences the decisions and to whom the case should be made are common questions which need to be addressed when lobbying those in power at local, national or European levels.

Chapter 8 addresses the need for construction companies to develop a more strategic approach to the communities in which they operate. An increased role and a more positive profile will do much to counter the confrontational image that has typified recent controversial motorway and other road schemes, and other developments. It may also provide a competitive edge and help to satisfy the increasing number of clients who themselves have a social remit. The chapter will stress the business benefits of an active community policy and successful implementation which involves construction companies being proactive in providing information on forthcoming/on-going projects, appointing community liaison engineers and establishing forums with local people and other interested parties.

Part Three identifies the role of corporate communications in addressing internal audiences through the use of effective internal communications techniques.

Chapter 9 focuses on *managing communications internally*. Managers and staff need to be informed of the business mission and objectives, and their own opinions and information are important in ensuring effective business strategies. More effective communication is a strategic tool for gaining employee commitment, raising morale and increasing productivity, improving quality standards and ensuring safety. In the fast-moving environment of construction, only companies that can respond quickly to market opportunities and changes to working practices will remain successful. The chapter views internal communication as part of the development of corporate culture, supportive of internal marketing and, at the operational level, vital to the success of the construction project. Tools and techniques are examined including induction packages for new recruits, company manuals, staff newspapers and verbal presentations to staff at all levels.

Part Four looks at communications management at the level of construction projects and how public relations must be viewed as an integral part of project management in construction.

Chapter 10 considers *the role of the manager in improving communications* with the stakeholders of construction projects. It examines the range of skills required on the development of project-level public relations strategies, and outlines the support required from the company's public relations function. The chapter stresses the need to support the internal construction team given the role of non-marketing managers and staff involved in pre-qualification presentations and interviews with the client team. More effective approaches are examined, including greater cross-functional co-operation and communication, and the sharing of information on clients and projects throughout the company; the need for clear, well researched and well rehearsed presentations is emphasised. The chapter closes with an examination of effective approaches to site event management.

Chapter 11 focuses on communicating safety in construction and, not surprisingly, is the biggest chapter of the book. In an industry with a notoriously bad record on accidents, fatalities and injuries, the chapter stresses the need for public relations to be part of the organisation's promotion of its safety policy. Such a strategy will ensure that employees, subcontractors, suppliers and the client team are well informed on safety matters. The chapter details a number of techniques including safety campaigns, the use of the press and media in promoting safety on site, and the involvement of the local community, especially schools, in safety campaigns.

Chapter 12 builds on the previous chapter on safety communication and looks at ways of managing crisis situations. A crisis management system is a major priority for any business and needs to be much broader than simply a matter of good communications. It has to be developed in conjunction with all functions of the business and with managers and staff at all levels. The focus is on defining 'crisis' in construction and on explaining how companies can plan for the management of unforeseen incidents. The chapter outlines the contents of the 'incident management plan', which will help contain the incident and hopefully stop it becoming a crisis.

Part Five provides concluding thoughts on the corporate communications issues raised in preceding chapters, and on the more effective management of public relations strategies in construction businesses and projects.

Part One

Chapter 1
Corporate Communications and Stakeholders

Introduction

This chapter focuses on the importance of stakeholder concepts to business success in the construction industry and the use of stakeholder identification and analysis as a tool for corporate planning and communication.

Business strategies are concerned with the orientation of the firm to its external environment and the manner in which it uses its resources to achieve success. An essential contribution to this success is made by approaching the business planning process proactively. To achieve this, the construction business is dependent on information about the environment, covering past and future trends and emerging issues and problems.

What has the concept of stakeholders got to do with public relations?

Stakeholder theory provides a formal mechanism for identifying those groups that have a stake in the organisation; that is, those who influence or are influenced by the business. This allows communication activities to focus on those stakeholders that are influential to the organisation's development and business success.

> 'A stakeholder in an organisation is (by definition) any group or individual who can affect, or is affected by achievement of the organisation's objectives.' (Freeman, 1985)

Stakeholders of the construction industry include clients, government, the City, suppliers or subcontractors. The use of stakeholder relations ties communication activity more closely to the corporate strategies of the organisation. There is also the possibility of creating a greater dialogue between the organisation and its stakeholders. This opens up lines of communication and allows the organisation to obtain more feedback on its operations. Appropriate information can be better targeted at specific stakeholders.

The stakeholders in construction

The concept of stakeholders has its roots in the late 1960s and 1970s. Business environments had been relatively stable up to this period but had become increasingly turbulent. In stable environments few surprises occur but the world was changing rapidly and the need to plan for change was becoming apparent.

An early approach to the stakeholder concept was called the 'production view' of the organisation. This suggested that the organisation need only satisfy its suppliers and customers. It took the view that these were the only external groups of which the company needed to be aware. This approach is shown in Fig. 1.1.

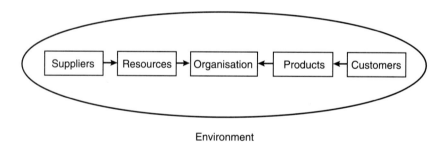

Environment

Fig. 1.1 The 'production view' of stakeholders (adapted from Freeman, 1985).

While this approach is probably adequate for owner-run businesses, the moment they expand they are brought further into contact with the external environment through increased economic activity. In a construction context, the 'production view' approach would mean that the construction company is only concerned with its suppliers, sub-contractors and customers.

The next view of stakeholders is provided through the 'managerial view' of the organisation. It stresses that management has a role in developing the stakeholder focus of the firm. It also introduces the importance of internal stakeholders, i.e. the shareholders and employees of the organisation. This view (shown in Fig. 1.2) regards the company as a resource conversion entity.

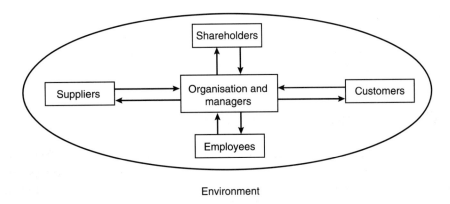

Fig. 1.2   The 'managerial view' of stakeholders (adapted from Freeman, 1985).

From Stanford Research Institute's work an early definition of stake-holders emerged as 'those groups without whose support the organisa-tion would cease to exist' (SRI, 1963). This definition fits very neatly into the managerial view of stakeholders illustrated in Fig. 1.2. This approach identifies the *primary* stakeholders of the firm and indicates the con-tribution they make to the organisation.

The primary stakeholders in construction

The shareholders of the business are primary stakeholders as they pro-vide the equity and take the risk by investing in the organisation. They have an intrinsic interest in how the organisation performs and can demand changes which directly influence the success of the company.

The customers of construction companies are interested in the products and services the organisation produces. Construction clients are increas-ingly adopting issues that mass consumers adopt. They are interested in achieving value for money, want the highest quality and increasingly look to the level of customer service that is provided. Companies that do not meet these criteria find their chances of follow-up work significantly reduced, and hence their market share suffers.

The suppliers of construction organisations, in the form of sub-contractors and material and equipment manufacturers and suppliers, influence the conversion process of the firm. They are an integral part of the service process of the company. Their performance and support are essential if they are to achieve the targets necessary.

In a similar way the employees influence the performance of the organisation. They contribute to the value creation within the company. A workforce that sees the importance of their contribution to the organisation is more likely to perform in an appropriate manner.

Ackoff (1974) takes the view that, by using an open systems perspective, many of an organisation's problems could be resolved with the support and interaction of stakeholders in the system. This is a rediscovery of Ansoff's (1965) original theory. Ackoff's approach adopts a co-operation philosophy where the organisation and its stakeholders plan together for the future. It is therefore based on the participation of and collaboration between the various stakeholders. An example in construction would be partnering arrangements between clients, consultants and contractors, initially on specific projects, but which could lead to longer term business relationships.

Secondary stakeholders

The stakeholder movement benefited from research carried out in the field of corporate social responsibility (CSR) (Post, 1981). The major contribution from this group is the application of stakeholder ideas to groups regarded as non-stakeholders. The majority of these groups had an adversarial relationship with the organisation. In the construction context, environmental groups and the disaffected community would form part of this collection of stakeholders. The CSR approach requires companies to be more responsive to external pressures and to adopt a positive change attitude.

Industries such as oil have attempted to take a more proactive approach by engaging secondary stakeholders. They attempt to engage environmental groups and gain feedback on their operations to mitigate possible negative reaction.

How can we manage our stakeholders?

The idea that stakeholders can influence or are influenced by the goals of the organisation indicates that we need to develop tools and techniques to ensure their inclusion in our communications strategies. There are a number of issues that we must understand.

(1) Who are our stakeholders in the organisation and what is their perceived stake?
(2) The processes and techniques the organisation will use to engage stakeholders and develop relationships: how do these fit into the organisation's overall business strategy?

(3) The 'transactions' that take place with the stakeholders: to what extent will these influence business decisions? How will the company address future negotiations with stakeholders?

Any organisation may be able to identify respective stakeholders but does not necessarily have an awareness of how they are to be engaged such that they form part of the business strategy.

Who are our stakeholders?

As we have already seen, construction companies can take the approach of splitting stakeholders into primary and secondary stakeholders (Ansoff, 1965). The positioning of these stakeholders is indicated by their proximity and relationship to the core business of the organisation. The primary stakeholders are those having a direct and economic impact on the organisation. These are identified as the:

- Owners/shareholders
- Suppliers
- Competitors
- Employees
- Customers.

Add this in to plan.

primary

These stakeholders (with the exception of the competitors) are critical to the very existence of the organisation. They are the traditional stakeholders of the company and have a role to play in the development of the business.

If one were to take the case of shareholders in publicly listed construction companies, then their influence on the development of the companies is obvious. Over the past two decades construction companies have expanded, contracted, been involved in mergers and acquisitions, and changed the way they operate. During this period the major equity holders have monitored the companies and their performance, and in certain cases supported activity and in other situations opposed proposals. Senior management has often been held accountable for performance, and consequently there has been a turnover of management. The direction that an organisation takes is monitored and the major equity holders can influence the direction they take.

Competitors pursue strategies which seek to gain advantage over us. This is not only in the business context of winning more work, or in a technological sense but also in the social and political arenas. Competitors may want to increase the level of their influence on policy. A demonstration of this is the number of firms that seek to redefine themselves as 'non-adversarial' companies. This may be in response to client pressure. Less adversarial arrangements in the construction industry enable organisations to influence social and political views.

On the other hand the competitors grouped together form part of an industry block. By working together they can alter perceptions and form a stronger political lobby. This has been demonstrated by specialist sub-contractors who together have gone a long way towards removing and reducing punitive clauses in contracts. As individual companies this would not have been possible but as a group they form a strong lobby.

The secondary group of stakeholders comprises those individuals, groups and organisations that are not directly related to the core business of the organisation. They include groups such as government, local authorities, unions, local communities, political parties, consumer groups, etc. The diversity and potential influence of these groups suggest that secondary stakeholders can exercise the same level of influence on the development of the organisation. Many of the secondary stakeholders could have a significant influence on the company, particularly through the use of legislation.

All levels of government can often exert considerable influence on organisations. The introduction of compulsory competitive tendering offered a host of new opportunities to firms but the imposition of tougher environmental laws reduced development potential. The move towards private financing of public sector projects offers considerable opportunities to some organisations but also excludes large groups of companies.

The impact of secondary stakeholders in construction goes beyond the realms of government. The construction industry has a direct impact on the environment through its processes and end-products. Construction processes by their nature create an interest in the community. The result has been the setting up and growth of groups varying from those with a 'not in my back yard' (NIMBY) stance, to organisations such as Friends of the Earth. Secondary stakeholders can take their stake seriously by attempting to change the direction of planned or ongoing construction activity. This may be in the form of active lobbying or more direct action targeted at the construction process.

Understanding the 'stake'

Our analysis has to go further than just categorising the types of stakeholder. There has to be an understanding of the potential influence of the particular stakeholders over time. The adoption of an assessment that only operates in the present will inevitably suffer from the impact of a dynamic and changing environment. This is particularly evident in the construction industry which is evolving and therefore being influenced by new trends and issues posed by new stakeholders.

We need to establish the interest, influence and potential impact of particular stakeholder groups.

Economic influence

The most obvious dimension is the economic influence on the organisation. This is where the stakeholders may have an influence on the profitability, cash flow or share price of the company. In turn the firm may have an influence on stakeholders through its activities, whereby it can influence market share, growth, prices, etc. Fortunes are often won and lost on the performance of a company. The suppliers, subcontractors and joint venture partners would fall into both categories. Customers will also have a major stake in determining the economic future of the organisation. This is particularly evident in construction where the market mechanism is dependent on clients needing construction work.

Technological influence

Another dimension for particular stakeholders may be the technological effects on the organisation. Particular stakeholders may have the ability to prevent or enable the company to get access to technology, equipment or skills. The influence of this type of stake can be potentially strong if the technology is highly specialised and the stakeholder has a monopoly on supply. The use of specialist services is very prevalent in the construction industry, and as organisations outsource more of their work the potential influence of highly skilled, technologically advanced subcontractors will grow. Changing technology can also add to the power of specialists.

Changing public opinion

Some stakeholders may have a social influence on the organisation, by altering its position in society, changing public opinion, or otherwise helping or hindering the company. There is the potential for social stakeholders to act as a rallying point for other stakeholders. Consumer groups and environmental activists fall into this category. The influence of stakeholders is highly dependent on its values, morals and ethics. The value system of the organisation and its stakeholders will define this relationship and will depend very much on how the leaders of the company view the role of their business.

The political agenda

Social issues normally end up on the political agenda. Political influence followed up by legislative capability can have a serious impact on the organisation. All forms of government have this type of influence and may choose to exercise it to a greater or lesser degree. In certain cases the firm may have the ability to influence the political process through the lobbying activity of various stakeholders. These stakeholders with political influence can redefine the potential impact of change on the firm. The political stake has to be carefully considered.

Our own management philosophies

The company leaders' management philosophies and the culture of the business will largely determine the systems, procedures and values adopted by the company. In many ways this has a major impact on the relationships that exist between the organisation and its internal stakeholders. These relationships create the internal culture of the company. If a company operates in an environment of trust and co-operation, then it is likely that this will be carried out externally. The culture of the company is a function of both its internal and external relationships.

A number of construction companies are looking at the way in which they operate and the extent to which culture is influencing external relationships. One major construction company even went so far as to appoint a Director of Culture to the Board.

The way in which an organisation operates internally will also have an influence on the nature of the relationship with external stakeholders. Japanese contractors specify that individuals wishing to work with them should take a non-adversarial approach with clients. Staff being recruited have to show that they can work in a non-adversarial manner to fit within the existing culture of the organisation.

The categorisation of the stakes allows us to understand the nature of the relationship between the stakeholder and the organisation and actively address any issues that arise. In our understanding of the need to communicate with our stakeholders, the stake, or cause and effect, in the relationship is important. We need to ask questions such as the following:

- Why should environmental groups or local communities oppose development on greenfield sites?
- What is their interest?
- What do they see as the effects of such a development?
- What stake do they have ?
- How can they influence the outcome?

If we find answers to these questionsm then it is possible to open up lines of communication and develop strategies for dealing with such opposition.

The stakeholder map

An important tool of stakeholder identification is stakeholder mapping. The construction of a generic stakeholder map is relatively straightforward, as shown in Fig. 1.3.

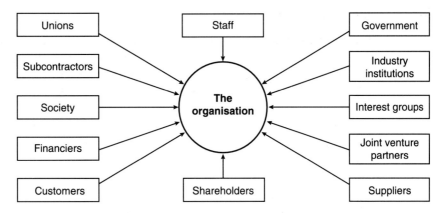

Fig. 1.3   A stakeholder map for construction organisations.

Building relationships

An important part of stakeholder analysis is determining who develops the relationship between the stakeholders and the organisation. In most cases this role is taken by the public relations or marketing functions within the company. These operate at either a corporate or departmental level to engage and develop relationships with stakeholders. However, the information and views gathered from these relationships are very often not fed back into the decision-making processes. The strategic and business development sides of the organisation do not see the importance of this contribution. Strategic decision making without due input from stakeholders could lead to the organisation not asking the right questions about itself.

Organisations are in a state of constant interaction with their stakeholders. This is most easily seen in the transactions that take place between the organisation and its stakeholders in the exchange of goods and services. The nature of these transactions goes further than the normal buying and selling of goods and services to activities that relate to other issues such as union negotiations, presentations to shareholders, government enquiries, etc. The comfortable seller–buyer nature of sta- keholding is changing rapidly and bringing new groups into focus. At the interaction level of stakeholder relationships there has to be constant updating of the relationships, or newer stakeholders will feel a sense of discontent.

It is commonly accepted that it takes a great deal of time to build trust and good relationships but a very short time to break them. The relationships that develop allow the organisation not only to have good sounding-boards but also to receive good feedback and support. The main issue is not to get complacent about the nature of the relationship or invest too heavily with certain groups of stakeholders to the exclusion of others. Tastes, opinions and support change.

A very good example of this is the change in attitude towards the natural environment, or 'green issues', over the last few decades. Environmental groups and their supporters were regarded as peripheral to construction projects but have gained a great deal of support and influence over this time. Companies that have engaged them and devel- oped relationships are in a stronger position to deal with the issues they raise than those that have ignored them. This scenario was illustrated in the petrochemical industry following the *Exxon-Valdez* incident. Exxon operated in isolation during this crisis. It could not build on experiences from its environmental stakeholders as it had no relationships with them. Its actions were misdirected and its responses inadequate (Smith, 1994).

The process of interaction between the organisation and its stake- holders is important as it defines the fit between its identification of stakeholders and mapping. The reality the company faces also has an influence. Information received through the process of engaging stake- holders has to be fed back into decisions or the efforts put into relation- ships are of no use. The objective of building stakeholder relations is to benefit from them.

Are our values different from those of our stakeholders?

One of the most difficult and most important tasks is the analysis of the organisation's own values and the manner in which they could possibly influence stakeholders. The individual value system is complex and is not always applicable to the organisation's value system. Values come in many shapes and forms. There are *aesthetic* values that provide views on what is beautiful or good. There are *social* values that identify what is good and just. There are *moral* values that identify goodness and evil. All these value systems have the potential to influence decision making on good or bad value systems. In order to help us sort our way through the complexity we can define values in two ways: those that are 'intrinsic' and those that are 'instrumental'.

Intrinsic values relate to those items and things that are valuable or good in themselves. They are pursued for their account and worth. Intrinsic values are pursued for their own worth unless there is conflict between them, for example the value of a Picasso over a Rembrandt. Another way of describing intrinsic value is that they represent the bottom line of life and its activity.

Instrumental values are means to intrinsic values. These are values that lead to something that has worth. The creative and artistic process that leads to the creation of a work of art may be considered as having intrinsic value. Instrumental values are not regarded as valuable in themselves but purely in the way they contribute to achievement of intrinsic value. Instrumental values abound in organisations as we concentrate more and more on the process of work rather than on the outcomes. This is typically a construction scenario. It is easy to concentrate on the instruments, for example the process of building a hospital or an office block, rather than the outcome, i.e. the benefits that the hospital or office block will have for those who will use it. This approach can have a fairly significant influence on how we approach our stakeholder relationships.

Many of the activities of an organisation on a day-to-day basis are intrinsic values in the organisation. In the same manner the process of building stakeholder relationships should help build intrinsic value for the organisation. Within this context, communication activities should contribute to the value to the company, and the process and actions are of purely instrumental value unless they deliver a desired outcome. This is something the construction industry should be aiming for.

Managing a conflict of values

There is a great deal of opportunity for conflict between the stakeholders and the organisation. This is highlighted where the value system of the organisation and its managers and employees differs. While it is not possible for the values of the people in the company and the organisation to be identical, there should be a degree of fit. Most importantly there should be a considerable degree of fit between the values of the senior management and the organisation, as they often provide the external face of the organisation. Conflicts have arisen in recent times over so-called 'fat cat' salaries.

There are also likely to be differences in the value system between the organisation and the external stakeholders. By having a clear organisation value system it should be possible to identify how these stakeholders' values differ from the company's values. Remedial steps may be taken if necessary. This could result in changes in the process used in relating to stakeholders and particularly impinge on what the organisation communicates.

Value analysis process

The flowchart in Fig. 1.4 indicates that in understanding stakeholders we recognise that there is potential for conflict.

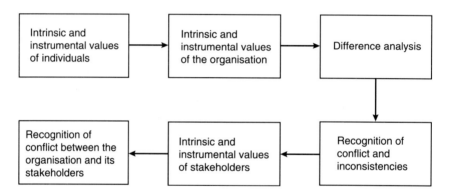

Fig. 1.4   Value analysis process.

It is possible to take a proactive stance to counter differences and objections. For example, on a development project that threatens woodland and habitat, construction companies are likely to encounter opposition to the project from local people and environmental groups. The company, as part of its value system, must recognise the need for a sound environmental approach to its activity and should attempt to engage secondary stakeholders as early in the process as possible. This will provide feedback to the company on likely opposition to its development. In the presentation of its proposal the company can include mitigation measures that reduce environmental impact. The manner in which the mitigation is presented will be designed to bridge conflict and show an express understanding of the stake of the environmental groups. This approach will not necessarily satisfy all groups involved but will go some way towards gaining support. A proactive approach to stakeholders would result from recognition of the differences in the value system of the company and of the stakeholders.

Summary points

This chapter has explored the concept of stakeholder theory and its relevance to corporate communications and business success. It has raised a number of ideas and issues that are important to the firm:

- Stakeholder theory helps us identify those groups that have an influence on the firm and those that we influence.
- Not all stakeholders are equally important. There are those that have a primary influence on business decisions and those that have a secondary influence.
- The importance of stakeholders is not static and may change over time.
- Adopting a stakeholder approach to business ensures that the firm questions its own value system to achieve a better fit with the environment in which it is operating.
- A stakeholder approach means that appropriate relationship-building and communication are directed at particular stakeholders. The techniques and approaches used for each group of stakeholders will therefore be more appropriate.

Chapter 2
Corporate Identity – Not Just Pretty Colours and Logos!

Introduction

A strong corporate identity or 'house style' in this very high profile industry can be a two-edged sword. The nature of the business enables a company to promote its corporate identity to an extremely wide audience through its everyday, highly visible construction activities – whether housebuilding, commercial building or civil engineering. Conversely, this high profile is usually associated with traffic cones, congestion, disruption to pedestrians, dirt, noise, dust and other forms of environmental pollution. Thus, the more strongly a company brandishes its logo, colours and other symbols of its identity, the better it needs to perform in terms of associating a positive image with that identity.

The purpose of this chapter is to assess how best to use corporate identity within the construction industry, how best to protect and enhance that identity and how to reduce negative impacts upon a company with a strong identity. However, first a brief examination of the fundamental elements of a corporate identity will help paint a background to the subsequent review of methods and implementation.

A vital strategic tool

It is necessary at the outset to state categorically that corporate identity is a vital strategic tool that plays an often crucial role in a business's success. It is also necessary to remind ourselves that corporate identity is not simply a logo, set colours, a typeface and flags, but about an overall image that is portrayed by these elements; there is no substitute for a company's good culture, service, management and performance. Most importantly, the identity or image of an organisation is most strongly identified with its people and their personality and culture as an association with the name of the company. This includes the visual expression of a company.

For a construction company, identity is expressed by everything it owns and operates: the livery of its plant, vehicles, site hoardings, cabins, flags, site signboards and even helmet stickers and high visibility jackets. But it is also expressed through the behaviour of its people who work surrounded by that visual expression: the wolf-whistling, unhelmeted, swearing, discourteous site worker will, particularly when so visually identified with a contractor, give a pretty tarnished image of that company to any passing public observers.

A further expression of identity is reflected in what a company communicates to its own employees and the outside world. This includes not only all its letters, faxes, e-mail, compliments slips and so on, but also its publicity material in the form of brochures, advertising, press releases, exhibitions, staff newspapers and annual reports and accounts.

A successful corporate identity that reinforces and strengthens a company's image, particularly in the construction industry, has to be well managed. It cannot be cosmetic, but needs to be controlled, steered and co-ordinated across the full range of expressions available to an organisation, backed up by real behaviour and personality characteristics that will reflect a positive image.

A need to be different from the competition

A strong, positive corporate identity can, if so managed, develop the unique corporate culture that is so important in differentiating leading construction companies, and then it can provide a real focus for creating internal loyalty. This combination can be a catalyst for improving business success.

The construction business is a small, often incestuous world with the construction engineering profession being a common denominator. To differentiate between the main contractors or consulting engineers, a sort of tribalism has arisen, the corporate colours being the most powerful expression of a construction site's sense of belonging. Like the differentiation between Manchester United and Manchester City, the strong colours of Tarmac differentiate between them and, say, Balfour Beatty. The 'team' and supporters are linked to the colours, the flags, the hoardings. So with the engineer's helmet! Being part of the team and identified so clearly and visually with the organisation and all its activities is often a significant element in retaining and enhancing staff morale as well as for staff retention and even recruitment.

Construction is basically a service industry; it is about people and their skills and how they interact with clients as well as how well they perform in their construction activities. Their contribution to the team needs to be recognised by themselves since the best staff usually value having a good and professional image of themselves. If they feel good in being associated so visibly with an organisation and at the same time know that their own input is significant for the well-being of the whole team, then this 'belonging' and pride can be enhanced. This in turn will generate greater commitment, loyalty and dedication.

However, if those same dedicated and loyal team players find that whenever they visit a construction site within the organisation the cabins are always dirty, that noticeboards are packed with the curled-up remnants of out-of-date memoranda and instructions, that the receptionists and site staff are always rude or abrasive, then their perceptions of the organisation, their own image and their own sense of belonging will be decidedly tarnished. Moreover, if they visit the area offices or even head office and find a shabby reception, a lack of professional care and attention to the building itself, and less than warm and welcoming receptionists, then those people's self-esteem as part of that team will do nothing to encourage further commitment and dedication.

Thus the internal promulgation, enhancement and strengthening of a corporate identity is every bit as important as the external enhancement of an organisation's image. Once the staff and employees are dedicated to that image and have pride in being a part of it, then they themselves will in turn become the best promoters of the image externally. In a people-orientated environment, the personality and culture of the people representing a company are critical elements in forming the image perceived by others outside that organisation. No amount of corporate advertising and mailshots will substitute for the good behaviour and image portrayed by the people employed by a construction company if it is to have a good reputation.

What do we want to achieve?

The objectives of any corporate identity programme within the construction industry are quite straightforward. They are to differentiate one company from another in a positive way. They are to create an external, visual appearance that reinforces an existing, positive image. Through consistent and comprehensive application, the corporate identity's objectives will help create a strong sense of belonging and a culture that is self-generating and encourages protection from within. The willingness of staff to protect that image will then lead to better performance in the work environment and to a greater willingness to show customer focus and quality of performance. This completes the quality circle of self-generating image enhancement, both internally and externally.

Protecting the quality circle

Another prime objective of implementing a strong corporate identity in this context is to protect the corporate identity quality circle; to ensure that no damage is done to the image of the organisation that could reverse that circle. Any such reversal, through a company's bad performance, through lax implementation of a house style, or through simple apathy, will quickly lead to a disintegration of a company's image. Apathy leads to lack of regard and of team effort, disillusionment, lack of commitment and a downward spiral of lack of self-respect and corporate image. These must be avoided.

A catalyst for change

In different circumstances, corporate identity can be a catalyst for change. Whether this is a deliberate refocusing of a company's objectives, strategy and mission or a more fundamental change resulting from merger or takeover, the corporate identity can be a major tool or galvanising instrument in bringing change to an organisation. That change means powerful and often painful internal change, not only to staff outlook and culture, but also to the image of a company as viewed by clients, investment analysts, consultants, the public and professional peers. Thus a major objective of corporate identity could involve implementing change.

A corporate identity's objective can be the most effective means of encouraging and promoting cohesion for disparate elements or subsidiaries of a large group or corporation of operating companies. As construction companies become larger, 'global' groupings across national and international boundaries, so a greater need for a co-ordinated image is required. In particular, those organisations growing through acquisitions, both domestic and international, need to reinforce amongst those acquired a sense of 'family', of belonging, despite the size of the acquisitive new owner. This does not necessarily involve the removal of the name, reputation or culture of the acquired company; it often involves enhancing that already established image but linking it in visually with its new parent group and its new sister companies, wherever they are based.

Despite these more proactive objectives, the basic aim of any organisation in establishing and strengthening its corporate identity programme is purely one of branding. A strong identity is just that: a brand, an image.

The public should be able to instantly recognise a Costain site, an Alfred McAlpine site, a Bovis site and a Birse site. Even with their new corporate identities, Balfour Beatty and Tarmac have quickly established their new branding. Recognition is very important to any organisation. It is also very important to employees, to clients and to others of influence in contract procurement. Familiarity breeds contentment amongst a wide audience. The bottom line is to create familiarity with an organisation among all observers.

Evolution

Three different scenarios of recent vintage serve well to demonstrate the role of corporate identity as a catalyst for change. The first is simply an updating to reinforce an existing, strong identity. The second shows a completely new identity to reflect a change of culture and mission. The third shows the use of corporate identity to bring cohesion to companies acquired by a new parent.

Keeping up with the technology

The first example can be used to illustrate, in purely practical terms, some of the scenarios painted where identity must reflect image. It involves a company that, after over 20 years of enforcing and enhancing its corporate identity throughout Europe and overseas, had been overtaken by technology. Computers and PC fonts, typefaces and applications had left its image looking tired and a little boring. However, the fundamental colours, logo and most other manifestations of its corporate identity were still strong and *sufficiently different to differentiate positively*.

This company, based in the Netherlands, had grown rapidly during the 1990s. With the majority of acquisitions being made in the UK and Germany, it found its various house-style elements being questioned by its new colleagues. The use of Prestige Elite as a typeface for all correspondence had been the firmly enforced style as it made it clear that such a letter or memorandum was 'typed' or personal, rather than a computer-generated and duplicated circular. Elsewhere, secretaries armed with PCs, be they equipped with Word, WordPerfect or whatever other software, could experiment with much more lively, attractive typefaces than the boring, typewriter-age Prestige Elite.

Similarly, the rigorous application of lower case lettering to *everything*, including company names, even the names of employees on business cards and proper names for addresses, streets, towns, cities and even countries became the source of increasing opposition among group companies.

As a result, the group public relations function established a house style committee with representatives from the parent group, a UK company, a

German company and two international subsidiaries. This committee met several times to identify key areas of opposition and to seek areas for improvement. However, by agreement of *all*, the corporate colours and the logo were sacrosanct, being so strong and recognisable throughout Europe and overseas.

The result was a significant but not too traumatic updating of the corporate identity. This 'tweaking' in itself gave a wonderful opportunity to go out and reinforce the whole corporate identity throughout the group and to tighten up a regime that, after so many years, had become rather out of date, pedantic and predictable. It also gave an opportunity for each group-company public relations practitioner to remind employees about his or her responsibility for implementing the rules and regulations appertaining to the elements of corporate identity: an important aspect in its own right.

The exercise culminated in a series of in-house corporate identity seminars being given by every public relations manager to relevant employees in every group company throughout Europe and overseas, backed up by a newly produced house style guide manual and an accompanying video from the group corporate public relations department. Both of the latter were produced in English, Dutch and German to help personalise the messages where possible.

The chairman of the group company endorsed the following introduction to this video and manual, both of which were aimed at ultimately reaching all 25 000 employees (specific mention of the company has been deleted for the sake of anonymity).

'The strength of our image

Our Group distinguishes itself from other construction companies in a positive way. Based on the quality of our products and services, as well as our commitment to both the environment and people, the Group is a leader in its sector and is recognised and respected as such.

Our Group enhances this recognition through its house style, which includes the building blocks for a well prepared, uniform presentation.

Proper use of our logo and corporate colours strengthens the identity of each group company while illustrating its interrelationship with other group companies: the Group as a unit. Consistent application of the guidelines specifying how the Group corresponds with others and presents itself in other ways reinforces the image of the Group.

Consistent application of the house style not only promotes recognition, but also has an economic benefit. Guidelines and standardisation make it unnecessary for each of us to re-think the manner in which divisions of the Group present themselves.

Since the introduction of our house style in 1970, developments in the field of information technology, and the expansion of the Group outside the Netherlands, have necessitated changes to certain elements of the house style. But the power of our basic colours remains unchanged.

Our task is to strengthen the familiar 'unity in diversity' and to reinforce the quality of our activities. This house style guide – including examples and instructions – is intended to help you to achieve this. It contains the guidelines to which we must all adhere.'

Following this endorsement from the most senior level within the group, the manual then reinforces the fundamental building blocks of the house style with the statement: '*The logo, the colours, and the font are the most important elements of our house style. Used properly and consistently they create – simple as they are – the basis for a powerful and distinctive Group image.*'

As with most corporate identity manuals or guidelines, this house style guide then proceeds to list all the technical applications of the main elements including letterheads, business cards, compliments slips, envelopes, fax header sheets, memorandum paper, and the rest, identifying font sizes, margins, word wraps, positioning of dates, headings, references, signatures and the like. Plant, hoarding, signage, flags and any other items connected with everyday work practice and application are covered.

To help reinforce the introduction of the new or improved elements of the house style, articles were published in every staff newspaper encouraging compliance and consistency. The first input came from the main corporate or group public relations function and heralded the 'imminent' introduction of the revitalised house style. This article lent authority to the forthcoming dissemination of the more detailed changes to the corporate identity and helped individual operating companies to lay down in advance the ground rules for the impending seminars, video and guideline distribution.

Whilst describing the mechanics of the revitalised elements, the 'corporate' messages also included this explanation:

'The company house style is almost a quarter of a century old, but this does not automatically mean it requires renewing. Indeed, the most important elements have lost none of their impact, and for many applications (particularly involving equipment and plant) the style will remain as it was. So no bulldozer, truck or tower crane will be immediately sprayed over.

However, due to developments in information technology (PC and laser printer) as well as the growth of the Group outside the Netherlands, all stationery could benefit from a slight 'face lift', which is why it is now being modified.

The most important rules regarding the amended company house style are explained in the new "house style guide".'

Nevertheless, even at this 'corporate' level, some very practical, business-like explanations were made to reinforce the rationale behind certain elements of change. For instance, in the case of the business card:

'The business card is where the company house style is most often seen. Although the existing card does make a distinctive impact, it still has disadvantages: the company logo in relief is lost when it is copied and it also makes the card unsuitable for double-sided printing 'English/ Arabic or English/Russian, for example'. For this reason, the logo will now be in the house colours.'

A few months later another article was published in the staff newspaper of individual operating companies; this time written specifically for and about the individual company concerned, customising or more finely targeting the group or corporate message of the previous article. More-over, it enabled the individual public relations managers at operating company level to link the new house style elements to their own sub-sidiaries, divisions and area organisations. Finally, it gave an opportunity to instruct *all* staff on the exact implementation strategies and timetable for the new changes.

'GROUP GETS ITS HOUSE STYLE IN ORDER

As was mentioned in the last issue of the staff newspaper, the Group is modifying some elements of its house style, or corporate identity.

The most important elements however, such as the colour format, the logos, the use of the set typeface for all printed material and the visual identity of all plant, remain the same.

The most obvious changes are the use of upper and lower case in letterheads, business cards and other printed material, and the increasing use of the group colours on some documentation. These – together with some text position changes – affect all Area, Division and Departmental letterheads, compliments slips, memorandum sheets, business cards, fax sheets, continuation sheets and, where appropriate, envelopes. Other printed material such as forms will change in future, but not immediately.

The main difference, however, is the use of Helvetica typeface for all correspondence.

Although some Departments and Areas have already adopted Helvetica, others have kept to the official Prestige Elite typeface, while some seem to use whatever they fancy!

From now on, all correspondence should be in Helvetica 10 point – the use of Arial or other similar typefaces is not accepted. If your printers have any problems with Helvetica, the information technology manager at Head Office will help resolve the technical problem.

If Helvetica 8 and 10 point are not available (or if a typewriter is being used), Prestige Elite should be selected.

All of these house style changes have come from a Group committee comprising the Public Relations managers of five group operating companies.

Briefing
During the next few weeks, the Head of Public Relations will be briefing Area Offices on the new house style and will be issuing copies of the 'House Style Guide' to all secretarial and clerical staff. This outlines the key elements. In the meantime, any questions can be addressed to him at Head Office.

Everyone's co-operation is sought in helping the smooth implementation of these style changes.

Good luck!'

The implementation was, ultimately, highly successful. The house style was imposed for the vast majority of stationery items and implemented rigorously across Europe by all group companies. Some had their problems, particularly in the context of computer technology, with fonts and typesizes causing problems for computer systems in the Netherlands, UK and Germany. Nevertheless, the attitude of 'any IT problems *can* be solved!' held sway, no matter how many complaints there were about computer problems.

Repositioning the brand identity

Tarmac undertook a major change to its corporate identity in April 1996. Explaining the changes, Chief Executive Sir Neville Simms was quoted in *New Civil Engineer*, (1996) as saying: 'Our new image is part of the overall restructuring process that has been under way for some while. This includes the Wimpey asset swap.'

The dropping of the old seven Ts that had been used by Tarmac since 1964 and the adoption of a new logo and completely different corporate colours was a radical, bold move. Not only was the new logo designed to reflect the changes being undertaken structurally, as mentioned by Sir Neville, but it was intended to present a softer, less aggressive image than the seven Ts.

The new branding, created by design consultants Sampson Tyrrell Enterprise, was designed to bolster efforts to change the company's image from 'adversarial contractor to construction services conglomerate more interested in partnership than subbie-bashing' (*New Civil Engineer*, 1996). This aim was put in more gentle terms within Tarmac's corporate identity manual which contains, as part of its introduction, statements of encouragement for all employees to enhance the group's 'world class' image:

'Why does our identity matter?

Our ultimate aim is to be seen as world class in whatever we do. How we appear to the outside world is an important element in reinforcing a world class image.

In these guidelines we have taken the values which capture the spirit of Tarmac and reflected them in a comprehensive design system.

To be thought of as world class we must, amongst other things, look the part.'

The manual not only gives the full technical content of the new corporate identity, its logo, colour systems, typefaces, and so on, but also exhorts the reader to think about the new identity; to associate with the new ideals, culture and objectives. For example, three of the new buzz words of the construction industry today are picked out and put into the context of the Tarmac corporate identity:

Customer focus

Designing with the customer in mind.
Building confidence, being accessible.

Quality

Quality in everything we do. Being
passionate about our most valuable
asset; our identity, exceeding expectations.

Innovation

Looking at old practices and problems in
a new way. For example, using new ordering
procedures, materials and designs, to give
a fresh, innovative look to the Group.

Although the following statement in *New Civil Engineer* (1996) elaborates upon Tarmac's objectives, it could equally apply to every large construction company in its sentiments. Image is increasingly important in the business world, and construction companies worldwide are having to extricate themselves from the 'muddy boot' and 'builder's bum' image and create a much cleaner, more responsible and professional feel that must be appealing to new partners in the industry. So, whilst being somewhat mischievous in its summary here, the magazine actually gets to the roots of the objectives in this exercise:

> 'Tarmac has a new logo. But far more than this, it is pushing hard to present a whole new corporate identity to dispel the old macho, adversarial, litigious, subbie-bashing, late paying, client unfriendly image that has for so long surrounded contractors. For Tarmac doesn't just build motorways and power stations any more; it manages property, designs structures, and even, strangely, designs uniforms for hospital staff.'

Mergers and takeovers

A press release issued by HBG Construction Limited on 7 April 1997 outlining the complete rebranding of its three building companies came only three months after the company's foundation following the acquisition of Higgs and Hill. The rebranding or establishment of a new corporate identity was seen as a fundamental part of building up the new organisation, and ably demonstrates how powerful a tool this branding can be. The press release stated:

> 'HBG Construction today announced a major step forward in the development of its UK building business. A programme of restructuring and rebranding, designed to make the company a world class competitor in the UK, is to begin with immediate effect.
> HBG Construction, owner of Kyle Stewart, GA and, from January of this year, Higgs and Hill, is already the fifth largest building contractor in the UK, with a combined projected turnover for 1997 in excess of £500m. However, effective exploitation of the combined balance sheet as well as skills and experience of all the companies to take the company forward requires significant reorganisation.
> The most obvious manifestation of this will be that, from April 7, the companies within HBG Construction will incorporate the group name into their own, so that they will trade as HBG Kyle Stewart, HBG GA and HBG Higgs and Hill.'

Also part of the same press release was a quotation from the Chief Executive, Adrian Franklin, who once again reinforced the need for a strong, cohesive visual identity to bring the disparate parts together as a powerful force within the industry. He stated:

'The re-structuring of HBG Construction's operations within the UK is a clear indication of our commitment to the UK construction industry and our belief in its underlying strength. The three companies that make up HBG Construction have established strong reputations for producing quality work over many years. That is why HBG Construction is already a major player in the UK building industry.

But to move forward further and to exploit the opportunities which we believe are available, the three companies need to become bigger than the sum of their parts. Now we have to take the company on to the next stage of its development, making it a truly world class competitor in the UK, able to bid for and win the most prestigious and profitable projects in the country.'

Coinciding with this announcement, all three operating divisions began implementing their new house styles. The quickest and most obvious manifestation was the immediate change to how receptionists fielded telephone calls: 'Good Morning, HBG Kyle Stewart, how may I help you?' This was rapidly followed by all-site signage, flags, vehicles, helmet stickers and other elements falling in line with the new style.

The new Site Identity Manual, which was distributed throughout the organisation, gave the three plant and equipment departments very detailed guidance on implementation and instructions on every visual element available to them. The same document exhorted everyone's enthusiasm for the whole project. Setting out its objective 'to outline the potential branding opportunities available to the company on site', the manual's introduction stated:

'Corporate identity on site plays a crucial role in making the names of the HBG Construction group companies known universally, reinforcing their membership of the worldwide organisation HBG, Hollandsche Beton Groep nv, and HBG Construction Limited in the United Kingdom.

The aim is to create a strong branded image through the use of the company colours and logos, to ensure that a site can be instantly recognised as an HBG Construction project.

It is essential that the strict guidelines laid down are meticulously observed.

This comprehensive house style manual has been produced to enable you to fulfil all aspects of site corporate identity.

- Logos and their use
- Type and colours
- Site signage and banners
- Composite signboards and information signs
- Hoardings, fencing and site accommodation
- Plastic sheeting and debris netting
- Plant and vehicles
- Tower cranes and flags
- Helmets and site clothing.'

Making sure corporate identity works

Pick up any corporate identity handbook, manual or other guidance documentation produced for a company's image implementation and we find exhortations to enforce rigidly each element to the letter. These are directed at every employee and range from laying out a letter in a certain format to the size and positioning of the helmet badge. But since every single employee *can* impact both positively and negatively on a company's image, it is important that the house rules or guidelines be promulgated far and wide.

When a new or revised corporate identity is produced, there are many opportunities to get the message across to staff (as shown in the three examples above). However, it is after the main thrust of change that implementation becomes more difficult; it becomes everyday, familiar and therefore relaxed. Gradually a blind eye can be turned towards the oddly laid-out fax sheets, the sticking of a logo on the wrong coloured background, the slightly different shades of corporate colours on that flag or site hoarding.

Only constant monitoring and exhortation will ensure a consistent and comprehensive corporate identity. Should there be more than one set of house style guidelines: one for the full technical implementation tool kit giving every finite element the full specification/coverage and another, more simple, version that is issued to every new member of staff upon induction and to every new site when it is established or mobilising?

For the more senior staff members, this can be reinforced through the official company staff manual that sets out all staff responsibilities and rules and regulations. One such manual specifically includes its corporate identity element within the public relations section and states:

'Consistent application of the house style promotes recognition, saves unnecessary work in laying out documents and helps the presentation of a consistent image. The image is one of unity within the diversity of a well managed and high quality group.

The Group gives clear, helpful and definitive guidelines on house style. The public relations department is responsible for overseeing and monitoring adherence to them. It is to be consulted whenever there is a possibility of departing from them . . . A House Style Guide booklet and a video are available from the public relations department. The guidelines given, including examples and instructions, must be adhered to. Any queries should be referred to the public relations manager.'

For general, everyday implementation of consistent house style or corporate identity elements, constant monitoring has to be undertaken by the public relations personnel. Equally important is the constant reminder to staff of the need for a consistent style and the admonishment of lapses as quickly as possible.

One of the most common causes of divergence from a corporate identity is through the liberal use of word processors/PCs. Not only does computer technology give reasons for not being able to adhere to guidelines but it is such a flexible tool that it positively encourages innovative style layouts, typefaces and icons.

Noticing that computer technology can rapidly become a dogmatic reason for *not* implementing a house style, some companies send out with their corporate identity manual a separate sheet identifying all potential IT problems that could be confronted by any PC operated throughout the organisation. This includes 'advice' such as changing the print drives, upgrading software, setting up macros, fitting filter screens and the like.

However, it is the more *laissez faire* PC operator that is the biggest danger to corporate identity. The plethora of typefaces and sizes, together with clip-art, allows for an amazing array of documentation to appear on site or on office noticeboards that have no relationship whatsoever to style guidelines.

In the construction industry, secretarial support is often, by the very nature of the business, temporary or short term. Every time a site mobilises and is set up with offices, a local secretary is recruited to support the construction operation. Confronted by a PC and a bunch of project engineers with little guidance as to expected house styles, any secretary can be expected to produce letters, memoranda and faxes in whatever typeface seems appropriate. Therefore a mechanism needs to be established from every area office whereby guidelines are automatically given to any new secretarial recruit to a site as it is established.

Whereas staff may be given a full induction programme that includes guidance for the main elements of house style or corporate identity, locally recruited secretarial staff may not be educated according to such parameters. They must be! There must be a mechanism whereby any new recruits are made aware of the basic elements of a company's corporate identity. All have access to computers and faxes. All *must* know how to address the house style to these outward elements of corporate communication.

The fax machine

The worst culprit in any company (construction industry or any other) is the use of the fax. The fax cover sheet is so easily produced and customised for every department or specialisation that it can very quickly become an established and, worst of all, accepted, culture to adapt individual fax cover sheets to identify different functions or departments. It is very difficult to control.

A recent example of a directive to control an increasingly diverse application of fax cover sheets follows. This was issued in response to the obvious disregarding of corporate identity guidelines and by rivalry in the creation of new, alternative designs for faxes. This memorandum was issued by the company's public relations manager:

'To: Divisional Directors, Area Managers, Heads of Department

Subject: **Fax Cover Sheets**

The fax has become the standard form of communication in business and, as such, is one of the main media through which the company and the group is perceived and its image portrayed. It is therefore vitally important that we have a consistent and common presentation of our fax cover sheets if we are to enhance and protect our image and House Style.

The enclosed examples show clearly that sites, area offices and depots are competing to have the most diverse style of fax cover sheet combining wild graphics and a plethora of typefaces and typesizes. The House Style Guide (copy enclosed) gives very clear instructions on how the fax should be presented, using photocopied letterhead paper and a Macro, Helvetica 8pt and 10pt typeface and general layout. I attach a sample giving the correct presentation of a fax.

Please can you ensure that with immediate effect all sites, depots and offices adhere to these guidelines. Any IT difficulties in this context should be addressed to the Information Technology Manager at Head Office and stationery questions addressed to the Office Services Manager. Any question of house style should be addressed to the under-signed,

Public Relations Manager.'

Summary points

- The more strongly a company promotes its distinct logo, colours and symbols, the more it needs to perform and to live up to the image it is communicating. Image and reality must coincide.
- Corporate identity is a vital strategic tool which is essential for business success. It must reflect the culture of the organisation and must be expressed in everything the company does and says.
- For a corporate identity to be more than cosmetic, to really support a business and its operations and to provide a competitive advantage, it needs to be controlled, steered and co-ordinated. It has to be backed up by real and positive changes in behaviour by all managers and staff.
- A strong corporate identity can differentiate the organisation in the minds of clients and create a sense of loyalty among the company's own staff.
- A pride in the identity will improve the services provided which will achieve success in marketing.
- The corporate identity quality circle is all about ensuring that, once created, the corporate brand is well protected and no damage is done as a result of lax implementation or apathy from members of staff.
- Corporate identity is a catalyst for change and can encourage and promote cohesion in increasingly larger and indeed global corporations, where there is a need to establish a sense of belonging.

In an article about the new corporate identity for British Airways when it removed the Union Flag from aircraft tailplanes and substituted it with 'bold images of the world', *PR Week* (1997) magazine stated:

'Changing corporate identity is a tricky business. It's hard to get the media to appreciate the finer details of corporate repositioning and even more difficult to get the punters to understand why a change of image should cost millions.'

On the other hand, and equally simple, is the description given by Tarmac for its 'identity toolkit':

'Certain basic elements form the toolkit from which the entire Tarmac visual identity is built.

To ensure consistent quality standards across all our companies throughout the world it is vitally important that the rules governing these basic elements are adhered to.'

Chapter 3
Corporate Communications Management: Objectives, Skills and Organisation

Introduction

This chapter will identify the main objectives and functions of public relations in the more effective management of corporate communications in construction organisations. It will define the relevant management skills required of those charged with carrying out public relations functions. It will also consider the different management options available in the organisation of a public relations function, namely either internally through the setting up of a separate department or as part of an existing marketing function. The chapter will also discuss the issues relating to the selection and use of outside consultants. We point out the need for a public relations strategy aimed at building positive business relationships with a variety of different stakeholders at corporate and project level. These strategies, tools and techniques will be examined in more detail in later sections of the book.

What do we want to do?

The main objectives for establishing effective corporate communications for any business are to establish and sustain positive relationships with a number of important or influential groups; to anticipate trends, issues and events; and to plan and manage responses to them.

Corporate communications needs to be viewed as an integral part of the overall strategic management of the construction business. Corporate communications objectives must therefore relate to the general corporate mission, aims and objectives, and must also be part of the 'shared values' of the business. An organisation needs to stress the importance of effective communication and the building of good, positive relationships, especially with customers and clients, as part of its overall corporate culture.

More specifically, corporate communications provides support to marketing activities, improves relations with investors, local communities, management and staff within the organisation, and maintains and develops a clear corporate identity and image for the organisation, internally and externally. An organisation's corporate communications programme should also be used to assist in broader industry initiatives to improve the general public image of construction.

Therefore, corporate communications objectives will broadly relate to the company's objectives with regard to a wide variety of stakeholders. Specific objectives will be explored in more detail in later chapters.

What challenges do we face from inside and outside the organisation?

As we outlined in Chapter 2 where we looked at the stakeholders of the business, it is essential for any business to analyse its environments. Gathering information on the different environments and corporate communications audiences may be conducted in house by the company's public relations staff or by outside agencies. However this is done, it is essential to find out as much as possible about the environment of the organisation, both internally (the company's own employees), and externally (its clients and customers, media, communities, etc.).

Information may be in the form of published data, for example from scanning newspapers, relevant magazines, the internet, etc. This may be a good starting point for establishing the background, but it is unlikely to provide an organisation with an edge over its competitors as these sources are accessible to all. The company may hire the services of a research company, but should also use its own people at all levels as sources of information on current and likely future developments. This increases a sense of involvement in corporate communications and provides unique information to the organisation which may give it a competitive advantage.

What choices are we faced with?

The processes and techniques selected by an organisation will relate to the objectives and outcomes desired. A balance needs to be achieved between what the company can afford to do and the most effective tools based on experience and research into their effectiveness. For example, are advertisements in trade magazines an effective method of promoting the company to its customers or would having a stand at an exhibition be more appropriate?

What managerial skills do we need?

Good public relations is vital in organisations faced with a turbulent and highly competitive environment such as construction. It requires companies to allocate the appropriate level of managerial and staff support and budgetary resources to ensure that it is effective. All managers need to develop the following essential skills:

- *Knowledge of the organisation*: the need to be intimately familiar with every aspect of the organisation, its work and its objectives.
- *Communication*: communicating ideas and suggestions for improvements in the way the company presents itself.
- *Presentation* in different forms: presenting to both internal and external audiences in a positive and persuasive manner.
- *Crisis management*: the organisation must establish its approach to crisis situations. Senior management will rely on corporate communications staff in particular, to take the pressure off them in handling the general public, media and other interested parties while they concentrate on resolving the immediate problems.
- *Advising*: managers need to be supportive of all other functions, and a culture of co-operation and information sharing should be encouraged.

Corporate communications skills needed by other senior management

Public relations requires the full support of senior management, who need to be fully involved in implementing the strategy. It may be necessary for senior staff to receive formal training in presentation skills, particularly for certain media, for example the press, and television and radio interviews.

How are we going to do it?

Public relations needs to be viewed as 'everybody's business' in the organisation. But it is essential to create a public relations department. The company needs to make a number of decisions about this.

- What will the corporate communications function do?
- Should corporate communications be part of an existing arrangement for marketing or a distinct department?

- Should the company employ external specialists, i.e. consultants and agents?

What will the corporate communications function do?

Sheldon Green (1994) identifies three functions: managerial, executive and mechanical.

(1) *Managerial functions* include creating, budgeting and controlling a corporate communications programme, monitoring the effectiveness of the programme and modifying it as necessary, in line with overall corporate objectives and strategies. These functions require a consistent dialogue with other functions and departments of the business to ensure integration of corporate communications into the whole organisation.

(2) *Executive functions* include the application of the corporate communications skills introduced earlier. For example, much of corporate communications activity requires the use of the written and spoken word. It also requires the commissioning of work from 'suppliers' such as designers, photographers, printers and other specialists. The requirement is for clear briefs and a knowledge of how these consultants work and how much things cost! Corporate communications requires a great deal of organisational skills, and an effective approach to project management is essential. The corporate communications function of the business, as we have seen, will be involved in pursuing a variety of different objectives and using a range of techniques. The programme will therefore be complex in terms of the logistical skills required. Given the specialist nature of much of corporate communications work, for example investor relations and crisis management, specialists may have to be employed. Corporate communications will need the executive skills to manage these different organisations so that they contribute to the overall success of the programme.

(3) *Mechanical functions* may be many and diverse according to the size of the department, its objectives and its overall programme of work. The following is a list of possible functions requiring a broad range of technical, organisational and managerial skills.

  - Maintenance and distribution of media lists. (Which publications are important to the organisation in terms of coverage of the business's activities?)

- Media monitoring. (How is the organisation being covered by newspapers, magazines, and news media generally?)
- Establishing and maintaining a library of all communications material used by the business. This may include photographic, video or film records.
- Establishing and maintaining files on all important corporate communications contacts, interest groups and channels of information.
- Organisation of events.
- Designing and printing corporate publications, for example annual reports and accounts, in-house newspapers, business cards, etc.
- Controlling budgets in relation to the agreed costs of the overall programme.

Part of marketing or a distinct department?

As will be demonstrated in Chapter 4, public relations management has a crucial role in supporting marketing, and indeed in many organisations it is here that public relations is positioned as a function of the business. It may be that, if corporate communications is positioned within marketing, it is seen more positively to be part of the commercial success of the organisation. However, there are a number of arguments about why corporate communications should be separate from any marketing or sales function.

First, the objectives of a comprehensive corporate communications programme may be much broader than simply creating business and making sales for the organisation. They extend to including the way the organisation communicates with, and relates to, a whole variety of audiences, and not just its customers or potential customers.

Secondly, public relations in its broadest sense includes the way the organisation communicates internally. It needs to be viewed as integral to, and serving all the activities of, the business and not any one department.

Thirdly, if public relations is to be taken seriously by senior management, it needs to be viewed as a distinct and vital element of the overall corporate strategy of the business. As a department in its own right, with a director at board level, then corporate communications may have more chance of asserting the need for the communications aspects of the firm's operations to be given serious consideration.

Do we need to get outside help?

The specialist function undertaken by consultants or agencies could include media relations, design and production of publicity materials, investor and shareholder relations, corporate hospitality and sponsorship.

In selecting a consultant or agent to carry out specific work, the firm needs to consider the following:

- Their reputation and past experience. Has the consultant or agent worked in the construction industry before? Have they worked for the competition?
- What standard of work have they achieved? Request examples of the consultant's work.
- Providing the consultant or agent with as much information as possible on the company and what it wants to achieve, in the early stages of development.
- Establishing a good working relationship.
- Evaluating the effectiveness of the consultant's/agent's work.

Have we achieved what we set out to achieve?

The success, or otherwise, of the corporate communications programme needs to be evaluated in terms of identifying whether objectives have been achieved. Expenses that have been incurred need to be justified, often to senior management who may be sceptical about the usefulness of the programme in the first place. Based on this evaluation future improvements can be made to the strategies adopted.

White (1991) indicates two possible ways of evaluating the success of corporate communications campaigns. First, some factor should be established at the start of the campaign, such as awareness of a particular issue. The same factor would be measured at the conclusion of the campaign and the two results compared. The difficulty here is that it may not be possible to directly link any change in awareness to the actual corporate communications techniques used. A second possibility is to follow a corporate communications effort from the organisation, through whatever medium is used, to the recipient. This becomes potentially a very expensive exercise in terms of management time and possibly money.

Evaluation of corporate communications activities is a highly subjective process and is riddled with potential inaccuracies. However, it may be that some form of professional judgement will be needed to assess the success or otherwise of particular corporate communications campaigns.

Decisions will have to be made as to whether the cost of monitoring the programme will be justified in terms of potential benefits.

Summary points

- It is important, from the outset, to establish what the company wants to do. The company needs to establish clear and measurable corporate communications objectives for internal and external publics.
- The corporate communications plan should support more direct marketing efforts by improving relations with customers.
- The plan should ensure financial stability by communicating more effectively with investors and the financial community.
- The plan should also engage the organisation in relating more positively to the local communities in which it operates and which it serves.
- The organisation needs to carefully analyse corporate communications environments, i.e. clients, employees, suppliers and subcontractors, shareholders, government, professional bodies, print and broadcast media. This may be developed through seeking customers', employees', suppliers' and subcontractors' views on the company.
- The information which the company gathers will be essential in enabling it to make choices of strategy and techniques, i.e. managing corporate communications in-house or through outside agencies and the most appropriate tools and techniques to be used.
- Those charged with the function of corporate communications will need to gain top leadership support and involvement, and to engage senior management in corporate communications activities. This may involve the development of an in-house corporate communications team and the selection of outside agencies and consultants for specific projects.
- Feedback will be essential. What have been the effects of corporate communications in terms of improved recognition and/or relationships with clients, investors, employees and communities? What effect has this had on commercial success?

Part Two

Chapter 4
Marketing Communications – Caring for the Client

Introduction

This chapter will focus on arguably the most important stakeholder of the construction business, the client. It will assert the need to develop better communications and build long-term relationships with clients and their appointed teams. We will examine the need for a comprehensive approach to client care which reorientates the construction team towards providing high quality services. Here we show how public relations can assist in communicating with the client team through more effective marketing communication techniques, particularly during pre-qualification processes.

Public relations in support of marketing

Public relations, as discussed earlier, is more than a function of marketing and sales. It is the function that assists in the more effective management of corporate communications with a wide variety of different publics or stakeholders of the firm. The most important stakeholders of any commercial organisation are its customers or clients. It is therefore essential that public relations activities aimed at customers work in increasing an organisation's direct marketing efforts.

What is marketing?

Marketing is a key factor in business success. The term marketing must be understood not in the old sense of making a sale ('selling') but in the new sense of satisfying customer needs. Today's companies face increased competition. Rewards go to those who can best understand customer needs, and deliver the greatest value to their target clients.

Marketing has been defined in a number of ways. Kotler and Armstrong (1993) define marketing as a social and managerial process by which individuals and groups obtain what they need and want through 'creating and exchanging products of value with others'.

According to the American Marketing Association's (1960) official definition, 'Marketing is the performance of business activities that direct the flow of goods and services from producer to consumer or user'. In 1985, the Association approved this definition: 'Marketing is the process of planning and executing the conception, pricing, promotion, and distribution of ideas, goods, and services to create exchanges that satisfy individual and organisational objectives.'

The Chartered Institute of Marketing's definition is that, 'marketing is the management process responsible for identifying, anticipating and satisfying customer requirements profitably' (Dibb *et al.*, 1994).

From the many and diverse range of definitions the following core concepts emerge:

- Marketing is about satisfying customers' needs, wants and demands. This means actively finding out what clients are looking for in the services provided.
- It is about providing products and services that are of value to the client. This means understanding the client's business and how they will benefit from our service and use the end-product, i.e. the building or structure.
- Marketing means ensuring value for money and the satisfaction of customers. This requires working closely with the client team and identifying ways of improving the services.
- Marketing is about building relationships and working in partnership with clients and their appointed advisors to ensure satisfaction.
- It means making a profit for the business – we hope!

Relationship marketing and partnering

Relationship marketing requires effective communication with a long-term focus. It requires organisations to develop trust and genuine reciprocation and direct approaches in establishing client feedback. Continuous improvement can only be achieved if a system of complaints processing is instituted.

Relationship marketing demands that organisations get close to the customer and their representatives. In construction that means, to varying degrees, entering into partnering arrangements. This demands an emphasis on orientating the organisation's culture – the fundamental values and behaviour of the people of the business – towards the client. The emphasis has to be on the development and training of the organisation's people in the way in which they relate to and communicate with each other and the outside world.

Relationship marketing is an appropriate approach to construction services because the exchange process between clients and other key parties to a construction project is highly interactive and exists over an extended period of time (Thompson, 1996). Examples of relationship marketing strategies include the long-standing business links between Marks & Spencer and Bovis, and partnerships between local authorities, central government, the European Union and banks for putting together financial packages for projects.

Overall corporate communication messages therefore need to be consistent with these more direct marketing strategies. In deciding on marketing communications plans there needs to be an awareness of the potential public relations consequences and possible benefits which may result (Sheldon Green, 1994).

The importance of delivering a quality service to the client in construction and beating the competition

The clients of construction are broadly seeking long-term relationships with contractors and consultants based on partnership between compatible organisational cultures (Morgan and Morgan, 1991; Fellows and Langford, 1993; Preece and Moodley, 1995). More demanding and discerning clients are increasingly looking for high quality of service. This service must be experienced from initial contact, throughout the phases of a project, through completion, hand-over and beyond.

Achieving a competitive advantage increasingly requires closer understanding and identification with clients. That means getting to know the client's business and the needs and expectations of their consultant advisors, i.e. architects, engineers, quantity surveyors, etc. It also demands internal cross-functional partnering, sharing of information and clear communication between all functions of the business. A change in the business culture, which reorientates the people of the business to serving the needs of the client, requires considerable executive level support.

Quality service is an essential factor for the success or failure of any business. It is a profit strategy because high quality service will increase first-purchase clients and encourage repeat purchase by existing clients. In the construction industry in particular it should decrease the need for remedial work (Baron and Harris, 1995; Barsky, 1995; Storback *et al.*, 1995).

Success or failure in the construction industry is partly based on knowing the needs and wants of the client so that it will be easier to satisfy those needs. To do so companies should identify those things about the services they provide which can satisfy their clients and, one hopes, keep them coming back for more.

Why may clients be dissatisfied with the services they receive?

The construction industry is diverse and often involves a variety of parties with different and conflicting interests. This variety of interests provides a fertile environment for conflict in the industry. Moreover, this problem will escalate with the traditional separation of design and construction. Furthermore, there are other factors which increase client dissatisfaction:

- The finished product does not meet the agreed quality specifications.
- The contractors focus on the technical aspects of the product rather than on elements of the service.
- Conflicts arise between the personalities and management style of the contractors and those of the client.

What does the client want?

Each client has different priorities and needs. Contractors need to establish what priorities are sought by the client. A project of high quality, handed over to the client, tells only half the story. The standard of service in providing the project may have been less than satisfactory, with major conflicts between the parties rather than a partnership culture focusing on satisfying the client's needs and wants. To ensure client loyalty and thus repeat purchase, contractors need to provide a high quality service throughout.

Firms need to know how their products and services, marketing communication efforts, managers and staff, facilities and corporate identity affect clients and their advisors. They also need to know how the competition is perceived.

A company needs to establish why the client continues to negotiate and place contracts with them. What are their priorities? Is this merely repeat business, and are they continuing to look for better service from the competition? Just how loyal is the client?

Contractors need to be able to manage word-of-mouth communication based on previous clients' experiences, and use them as independent testimony to the quality of the service provided. Clients will use such information along with their own experiences to benchmark against competitors. The only way to find out whether a company has met or indeed exceeded client expectations is by asking them. This should identify strengths and weaknesses in the service experienced, and will aid in improving processes in the future.

Caring for the client

The late 1980s saw an upsurge in the development of customer care initiatives. It was a period when organisations found it increasingly difficult to differentiate their products and services from those of their competitors. Consequently, excellent service to the customer became a way of seizing a competitive edge (Pollock Nisbet Partnership, 1994).

The concept of customer care has been used widely in the manufacturing, service and public sector industries. Woodruffe (1995) states that 'service organisations are particularly dependent on levels of customer care'. This is a consequence of their heavy reliance on the people of the business to deliver services to customers.

The privatisation of public-sector industry has created a new emphasis on the customer (Cook, 1992). Many utilities companies and major parts of the National Health Service (NHS) have undertaken quality and customer care programmes intended to reorientate traditionally hierarchical and paternalistic based cultures into commercial enterprise.

Traditionally, the emphasis an organisation places on providing a good service to its customers increases as the product it sells becomes less tangible (Cook, 1992). So, for example, when a customer chooses to purchase a washing machine, after-sales service is only one of the deciding factors. Where there is no physical product involved in a purchase, the importance of the quality of services increases. An example of this is choosing an accountant or a lawyer: personal service is a very important factor in customers' choice.

The philosophy – teaching us to smile!

Customer care is the whole philosophy of treating customers well and keeping them informed (Smith and Lewis, 1989). But it is also about the way the company looks after its own employees through its style of management within the company and its terms and conditions of employment, etc. An organisation orientated to caring for its customers and clients does not just 'teach its employees to smile' (Clutterbuck, 1988), but changes its fundamental approach to achieving standards of service quality. It covers every aspect of a company's operations, from the design of the product or service to how it is packaged, delivered and serviced. It is part of the fundamental culture within an organisation.

Customer and client care is about managing the complex series of relationships between customers, individual employees and the organi-sation (Clutterbuck, 1988). The primary focus is, however, the customers, and ensuring that the firm provides them with a pleasant friendly greeting, a positive and helpful attitude, a professional and accurate business transaction, an apology for any delay, a quick resolution to any problems and a sincere 'thank you' for their business (Pollock Nisbet Partnership, 1994).

The benefits

There could be a number of reasons why a company decides to instigate a customer care programme (Cook, 1992):

- To differentiate itself from the competition
- To improve its image in the eyes of the customer
- To maintain or improve its market share
- To improve profitability
- To increase customer satisfaction
- To improve staff morale
- To increase productivity
- To encourage employee participation.

Having a properly managed programme would create a reputation for being a caring and customer-orientated organisation. It would foster internal customer/supplier relationships and bring about continuous improvements to the operations of the firm.

The programme

For customer care programmes to be successful, they need to span the entire organisation. Clutterbuck and Kernaghan (1991) identify a typical customer care programme pattern that involves some or all of the following six elements:

(1) *Decide the objectives and structure of the programme.* This is where top management should establish what the programme is supposed to achieve and outline its own role in making that happen.
(2) *Audit the current situation.* The organisation attempts, usually through market research, to find out what customers think of the quality of service provided, both in absolute terms and *vis-à-vis* the customer. It also looks internally, asking employees what they con-sider would be most unsettling for the customer in doing business with their departments.

(3) *Planning the programme.* This is frequently carried out in workshops that also present useful opportunities for team-building and for some problem-solving that will provide visible initial success.

(4) *Defining policies and objectives.* This should occur at an early stage of planning and involves looking at the most obvious barriers to customer care.

(5) *Preparing the ground through internal marketing.* In general, the bigger the change in culture and behaviour required, the greater the cynicism the employees will exhibit. To overcome that cynicism, top management has to communicate its intentions strongly and with conviction.

(6) *Devise an appropriate training programme* aimed at all levels of management and staff, particularly those in direct contact with customers.

An example of the application of these techniques in practice is the case of McNicholas Construction (*Contract Journal*, 1998). This family-owned contracting organisation launched the 'McNicholas CARES' initiative which urged employees of the company to be more proactive. 'CARES' is a mnemonic from the words customer, awareness, respect, enterprise, excellence, enthusiasm and solution.

McNicholas formed a series of focus groups of customers to identify their needs. The aim was to try to get close to the customer and to give a better service, particularly to the firm's clients in the water industry. Of an annual turnover of £180 million in 1997, £9 million came from the water sector, notably Thames Water with whom the firm had a partnership arrangement. The firm was reported to be winning work in Europe due to its reputation for commitment to customer service.

Communications techniques to support high quality services and customer care

This section has looked at the need for a more effective approach to increasing client satisfaction, to retaining customers and to creating new business. It has considered the need to build relationships with end-user clients, professional consultants such as architects and engineers, and all those involved in the delivery of a quality service to the client. It has identified a need to develop a customer/client care philosophy which may be translated into a formalised company programme.

Effective corporate communications is essential from the early contacts a company may have with its clients and their professional advisors, through the pre-qualification processes and into the construction and post-construction phases.

Good employee relations (which will be detailed in a later chapter) will ensure that those who are selling the firm's products and services are supportive of the overall marketing objectives and are committed to satisfying the customer.

The construction project site is considered to be the shop window in the construction industry. Clients are increasingly concerned with reflective image. Corporate communications activities in construction will largely centre around activities that improve the image of the organisation at this level. Strategies will be detailed in more depth in a later chapter.

Principal marketing communication techniques

Marketing communication activities that an industrial firm may be involved in are manifold and diverse. The concept of a 'promotional mix' (Rossiter and Percy, 1987) is based on the premise that the communication objectives are achieved through a planned and co-ordinated combination of personal selling and presentations, media advertising in the business and trade press, public relations and publicity events and other activities. The promotional mix provides the main channels of communication used by a company to present its messages to customers (Nickels, 1984).

Marketing communications techniques which may be applied to support a client-orientated approach include (Jefkins, 1996):

- *Corporate identity programmes and schemes.* These were outlined in an earlier chapter and are fundamental in creating a positive brand image with clients. They include business cards, vehicle liveries, letter headings, promotional brochures and other sales literature – indeed anything that represents the company physically.
- *House journals, newsletters.* These can be aimed at the people charged with the marketing and presentation of the company to clients and potential clients. Effectively this means anybody who comes into contact with the client and their representatives, from the telephonist, to the site manager, to the managing director. The company may use such publications to support the team and reinforce any training provided.
- *Management meetings and conferences.* These are useful in bringing together the team and those engaged in meeting prospective clients. They can develop better lines of communication and encourage information sharing, particularly across larger organisations.
- *Annual report and accounts.* These are likely to be produced by a corporate communications/public relations function within the larger organisation. It is important that all managers and staff engaged in

Magazines/technical press

Magazines are similar to quality in a sense since they rely on feature articles to form the vast bulk of their information. The main difference lies in the fact that magazines are generally interested in only one subject. Consumer magazines are usually bought by the general public and so are available in newsagents around the country. The number of copies of trade magazines sold is usually much less, since only those in the industry who have some connection with the magazine will purchase it. Those who read a magazine are doing so because they are specifically interested in that particular subject; hence there is less chance of the article being ignored by the reader. This makes specialist publications a very efficient method of advertising an organisation's activities.

The key technical press for the UK covering construction, building and civil engineering comprises *New Civil Engineer, Building, Construction News* and *Contract Journal*. It is through these journals that the vast majority of construction industry personnel find out about contracts awarded, the latest equipment and techniques being used, financial performance, and so on. It is also through these publications that the movement of senior personnel and management is most easily monitored. Thus, publicity (preferably positive publicity) through the pages of these magazines is a key target for construction public relations staff.

There is also a whole plethora of additional trade magazines and journals that overlap the construction industry directly and include significant volumes of features and news copy relating to construction companies. The *Architect's Journal, Concrete* and *Construction Repair* are included in this sector, but other industry-specific publications directly relevant to, for example, the civil engineering business include *Dredging and Port Construction, Rail Bulletin, Water and Sewerage Journal, Lloyds List, Ports of Europe, Highways, Tunnels and Tunnelling*. All play a key role in publicising those specialist sectors of the business.

Television and radio

The importance of television and radio as means of broadcasting information cannot be underestimated. With the potential growth of television and its already highly influential effect on society, it is clear that an organisation which uses television effectively to broadcast its activities will gain a competitive advantage over those who do not.

However, in the construction industry, radio – particularly local radio – is far more important. Where construction works affect the infrastructure and transport, be they road works, road construction or railway construction, the public needs to be kept informed about delays, contra-flows

Electronic media

The internet is by far the least understood and yet potentially the most useful form of media outlet. As computers play a larger role in society, the need for communication via the information superhighway grows and grows. More people than ever before are using the World Wide Web to locate information and increase knowledge, and the number of available pages increases every day. The irreversible growth of the internet means that organisations who wish to stay ahead must try to advertise their services and provide genuine news and current company information in some form on this medium.

What communications techniques may we use?

There are many different ways in which an organisation's activities can be broadcast, and it is important to remember this when trying to contact various media institutions. Otherwise, the effectiveness of the media campaign will be drastically reduced.

Press releases

Press or news releases are probably the most common type of material sent to media agencies. If an organisation urgently wishes to publicise news concerning its activities, it must make use of this type of communication.

In light of the speed of publication of news releases, there are guidelines to follow if the article distributed is to be accepted by journalists. Ridgeway (1996) identifies five questions that need to be asked and answered in order to write an effective press release:

- Who?   (has made news)
- What?   (has happened)
- Why?   (is it considered important)
- Where?   (did the event occur)
- When?   (did it occur)

As with any form of communication, when writing news releases do not tell untruths or exaggerate facts that cannot be proved. It is a good wish to have more detailed information on hand about the event, in case journalists are interested in the article and would like to know more. It is therefore also imperative that a contact name, address and telephone number be included at the end of the release; a surprisingly large number of organisations forget to do this simple task.

In the construction industry, press releases are generally aimed at the technical press (see above) and/or the financial press and so the text needs to be geared accordingly. Releases to the technical press need to be that: technical. They must be short (preferably on one page), sharp and to the point. Generally the text should start with the name of the company issuing the release, and then, in the context of a contract award for example, it needs to include: name of contract; name of client; name of consulting engineer/architect; value of contract; duration of contract; and location of contract. There should then be a synopsis of the purpose of the scheme (i.e. completed project) and then the main construction details (e.g. '1.7 km of dual 7.3 m carriageway with two precast and in situ concrete overbridges, a cut-and-cover tunnel 250 m long and 450 m of retaining walls') and any particularly innovative or outstanding features.

Captioned photographs

A picture is worth a thousand words. Captioned photographs are similar to press releases, except that a picture to demonstrate the event or activity is included to enhance the importance. The rules of effective releases also apply to captioned photographs. Deciding what photograph to include in the caption is not as easy as it sounds. It must be relevant to the organisation's activities, and, more significantly, it must appear interesting or exciting.

Feature articles

Feature articles or editorials differ from releases as they are generally much longer. This gives the organisation more of a chance to sell itself and its activities than it would have in a brief press release.

In the construction press only rarely can companies submit their own features as the editors prefer to originate and research their own copy. In this case, the features editor needs to be contacted and offered a topic or site which contains some elements or aspects that would be particularly interesting to engineers; an angle needs to be established to convince the journalist that it would be a good story. For any site visit, it is also recommended that a highly detailed briefing document be produced giving all the essential facts and figures concerning the scheme. That will

Advertorials of a free editorial nature are sometimes available in the less specialist journals. From a control point of view this is advantageous, since the message is conveyed exactly as the organisation wishes. However, since it has been written by a member of the organisation, it may be seen by the reader to be simply a cheap form of advertising (Smith, 1994).

Credibility derives from third-party endorsement (Dibb *et al.*, 1994). If the journalist or some respected personality alters the article and gives their opinion on the organisation, the article is seen from an impartial viewpoint. The obvious disadvantage to this is that the chance of the article being misinterpreted increases. Feature articles take a relatively long time to prepare. They should not be used for trivial items of news, when a simple press releases would suffice. The number of outlets to which the article is sent should also be carefully considered, and reduced to maximise effectiveness.

Television/radio interviews

The huge potential for television and radio has already been discussed. There are opportunities to promote and publicise a company's present and future activities, although caution must be exercised. In such a high pressure environment, a strategy needs to be pursued if spokespersons for the organisation are to come across effectively in an interview.

The only people who should attend an interview to represent the organisation are those who are fully conversant with its activities. Usually this will be a senior director or manager. In the construction industry, for example, the project manager may be the ideal representative for an interview.

Preparation is vital. It is all too easy to assume that because the spokesperson possesses significant knowledge about a company, he or she will be able to answer all the questions asked. This is unwise, since interviewers will deliberately try to ask questions which show the good and bad points about the organisation's activities. If the negative questions can be anticipated, a response can be prepared which could turn a potential disaster into favourable publicity.

Press conferences

A press conference is an event arranged by an organisation where representatives from the media are invited to attend in order to receive a message or opinion from the organisation. Its most significant use is where other forms of communication cannot convey the message quickly enough. The question should always be asked as to whether the conference is really necessary. Factors which will answer this question will be:

- whether speed of communication is essential;
- whether the importance of the news justifies the event;
- whether the news is so complicated as to warrant demonstration by experts;
- whether a product is so new that it must be seen in person to be appreciated.

Preparing for a press conference is similar to preparing for a television or radio interview. Literature concerning the event should always be handed out to all those present at the conference. This ensures that if a speaker fails to convince the media of their intended message orally, the information in the written form can be used as a back-up. Follow-up contacts should also be available.

Advertising and sponsorship

Advertising means paying a media institution in order to display a chosen message. Sponsorship is similar; here an organisation's name is linked to some team, personality or event.

Strictly speaking, advertising and sponsorship are general responsibilities of the organisation's marketing department. If those responsible for media relations in an organisation are informed on its advertising and sponsorship activities, efforts can be made to maximise the potential publicity that such activities will produce.

How can positive media relations help in a crisis?

The use of media relations for everyday situations and special events has already been discussed. However, an even more important situation can arise where the effective use of the media in a crisis can make or break an organisation (see Chapter 12 on crisis/incident management). Organisations spend large amounts of time and money on promoting a professional and caring image. Even so, it is still possible that a corporate tragedy may occur which will tarnish its image beyond repair.

When a tragedy happens, people wish to be assured on three points:

- that the organisation has done everything it could have to prevent the tragedy;

of view as well as a financial one.

To survive a disastrous event, an organisation must actively involve the media to demonstrate these three key issues, and adopt a strategy for dealing with them (Black, 1995). This strategy is very similar to the one for day-to-day media relations, except that it will need to be implemented much more quickly than usual. Therefore, all staff must know their responsibilities regarding appropriate sections of the media for a given situation, so that if called upon to give information they are able to do so effectively. To be sure that the strategy will be successful, it is worth testing it on a mock crisis from time to time.

How can we evaluate the success of media relations strategies?

Employing staff or an external public relations consultancy to manage relationships costs money. As a result, it is imperative that organisations achieve the maximum amount of publicity for a given amount of resources. Evaluation is required to monitor the effectiveness of the strategies.

The most primitive approach is to measure 'column inches': the number of columns of press coverage of the organisation in local and national newspapers. Press-cutting agencies take this further: not only is the total amount of coverage monitored, but the relative impact of each article is noted. Articles published are given a rating by the agency, depending on the type of newspaper and location in it, the amount of third-party commentary, and other factors such as whether contact names and telephone numbers have been included to assist those readers who would like to enquire further. This type of evaluation is designed to measure the *quality* of the coverage rather than the *quantity*. Television and radio coverage can also be monitored using the 'column inches' approach, although 'time on air' will be the variable measured.

The whole process of selecting a suitable form of media outlet must be evaluated constantly to ensure that the groups who need to be contacted have been contacted. One way of achieving this is to obtain feedback from the marketing department, whose market research should suggest whether the coverage has had a beneficial impact on the desired groups. However, the beneficial impact should be awareness and improved reputation, rather than increased sales. The media relations function is not directly responsible for advertising or other marketing techniques, merely for conveying a positive corporate image.

Current situation in construction

The construction industry is aware that its relationship with the media are not very good and it would like to see the situation improved. Some of the main institutions, particularly the Institution of Civil Engineers (ICE), have prioritised positive public relations as a key objective, and likewise the 1997 'Year of Engineering Success' (YES) initiative was very successful in the regions.

Certain types of media outlet are more applicable to the construction industry than others. Trade publications are used frequently, and they will still be able to be used in the future. The use of local media is rare and should be increased. The national media are very seldom used. National newspapers and television are generally not overly interested in the construction industry, and this will continue to be the case until the industry develops a better relationship with the media. There is, however, scope for the very large contractors to use these media outlets more frequently than at present.

The internet is as yet vastly under-utilised in the construction industry, and it offers much potential for those who are willing to put an effort into using it. It provides free advertising, and will reduce the workloads of media representatives. However, a web site *must* be updated regularly – more so than a press release, as it can be accessed at any time, on any day. There is increasing use of the internet by a wide number of groups, including clients, consultants and potential recruits.

Evaluation of media strategies is not as common in construction organisations as it should be, with smaller companies often being the worst culprits. The techniques of evaluation are often basic, and new media outlets and techniques of communication are rarely spotted. This means that media strategies are not effectively improved, resulting in the potential publicity being drastically reduced.

Those responsible for media relations in construction organisations are increasingly drawn from the public relations profession. In the past, far too many people with a traditional construction background took on the role, and they were not completely aware of effective ways to deal with the media; the situation was worst in smaller companies. However, most reasonably sized construction companies now employ qualified public relations personnel to handle media relations or have resorted to employing outside public relations agencies. The addition of experienced public relations professionals is vital if successful media strategies are to be developed.

Media representatives are sceptical of journalists, and are partly correct in this belief. Journalists in mass media outlets often neglect the construction industry and do not give it the credit it deserves. A fairer

Summary points

- There is a need to establish clear and measurable objectives of any media campaign.
- The company needs to identify the target and intermediate audiences it wishes to reach. This will dictate the appropriate media source and techniques.
- Information and news provided to the media, editors and journalists should be in a user-friendly style.
- An effective report to the media will clearly identify who has made the news, what has happened, why it is considered to be important, and where and when the event occurred.
- Effective media relations will help in a crisis if appropriate relations have been developed with journalists.
- Efforts should be made to evaluate the success or failure of media campaigns to ensure that objectives are being achieved.

Chapter 6
Financial Relations – Money Talks

Introduction – a highly specialised area

The day-to-day activity of a construction organisation does not normally involve much in the way of financial communications. The communications functions in the company tend to shy away from this specialist area and instead use external sources. Most financial communication is with shareholders, bankers, institutional shareholders, investment analysts, financial editors and regulators. These are stakeholders that take a specialist and highly critical look at the organisation. A secondary group, clients of the construction industry, are interested in the financial performance of the companies they employ and take an interest in their financial communications.

Day-to-day financial communications are often restricted to the preparation of reports and accounts. This type of activity will be prevalent in both listed and unlisted organisations. In the case of listed companies, financial communications are likely to be more specialised, covering share issues, rights issues and take-overs, which tend to fall outside the normal communication functions.

The financial stability of an organisation has to be demonstrated and proven to would-be clients before it can successfully pre-qualify for any major contracts. The last thing a client wants is a contractor to be liquidated or be made bankrupt half-way through a contract. Thus it is important for the marketing and public relations functions to have readily available financial information to demonstrate the company's financial strengths and viability.

Why financial communications are different

Financial communication is very different from other areas of communication because what is being sold is very different. The 'product' is the company, and financial communications can have a dramatic influence

of changes in the financial fortunes of companies (Jefkins, 1994).

In a normal public relations strategy, maximum media exposure is encouraged. Financial public relations can either force shares upwards or create dramatic falls. It can lead to accusations of insider trading, or make analysts suspicious of the motives of the company. Financial journalists also tend to be more critical and analytical, and usually do not take communication at face value. The City is well known for the number of rumours that circulate, and some public relations exponents see this as a legitimate way of communicating.

Financial communication should be frank, clear and informative. Communication should take both a personal and impersonal approach: impersonal in the form of news releases and briefings to the business press; and personal with key institutional investors and important shareholders. The extent of the success or failure of the communication can increase or decrease the value of the shares or damage people's perception of the company.

It is not only listed companies that need to be aware of the financial messages that are sent out about them but also companies that are not on the Stock Exchange. Financial checks are conducted on companies and this may influence the ability of the organisation to be accepted on to tender lists, etc. This may not seem to be the classical approach to financial relations, but the manner in which communication is conducted with the company's bankers can influence the outcome of relations. Banks will scrutinise the performance of the company in even greater detail and provide information accordingly on the performance of the business.

Financial stakeholders

There are a variety of financial stakeholders with whom the company needs to build relations:

- *Shareholders:* are particularly important in all companies and should not be ignored. In listed companies they have voting rights and can attend annual general meetings. Moreover they can be vociferous critics of the organisation if its performance does not meet the desired standards. The larger the block of shares held, the more powerful the shareholder becomes, hence the importance of institutional share-holders as discussed later. Shareholders should not be treated lightly as their loyalty and support are essential when fighting off take-overs or similar activity. Annual accounts, interim financial statements and other forms of financial briefings are sent to shareholders. In certain cases shares are held by groups who would not only on the financial activity of the company but also on its other operational aspects. Ethical funds taking large stakes in organisations enhance reputations but equally expect higher ethical standards.

The role of shareholders is equally important in privately owned companies. The impact of shareholders is probably more dramatic in that a small group can influence decisions and bring about changes within the structure of the company. As a point of principle these shareholders should be kept appraised of the financial performance of the company.

- *Institutions:* these are holders of large quantities of shares in companies with a view to increasing the value of the funds and income streams. Typically they are investment trusts, unit trusts, insurance companies and pension funds. These organisations wield a great deal of influence through their holdings in companies. It should be remembered that they themselves are under scrutiny for their financial performance. As major investors in the business they need to be kept informed of any changes in projected profits or any aspect of the organisation that could influence value such as changes in senior management, restructuring or contract awards. Loyalty with this group is performance related and they will withdraw support if they perceive that the company no longer provides good value. Institutional investor support is critical in take-over battles. They are also becoming more vocal if they are unhappy with an organisation's behaviour, as is evident with opposition to excessive pay settlements of senior management of underperforming companies (Greenbury Committee, 1995).

- *Investment analysts:* prepare reports on public limited companies and make recommendations on their investment prospects. They analyse not only the performance of the individual company but also how it performs in relation to its industrial sector, its geographic spread of operations, and the activities of its competitors in the sector. The basis of the analysis is directly related to the information from which the analyst is working. It is in the best interests of the company to provide analysts with information to get a more informed analysis of its operations. The views of analysts can make a difference as to how the company is viewed as an investment. Positive views mean stability and possible improvements in the value of the business, while downward forecasts invariably mean a drop in confidence in the company.

vision have specialist commentators on finance and economics. Financial journalists tend to take an analyst's view when reporting on companies. Their reporting is targeted at a particular market and helps to generate opinion about companies. In an age in which information is transferred rapidly, breaking news can have a major impact on the financial outlook for companies.

As has been demonstrated earlier, it is important to generate personal relationships of trust and honesty with financial journalists. This usually pays dividends in that a journalist needs friends in the industry to bounce ideas off and to assess broader situations. This is a chance to raise your company profile.

- *Banks:* have an inherent interest in the financial communications that a company makes. As lenders to companies, banks are keen to monitor any change in the risk of lending to specific companies. They closely follow reporting and analysis of companies and make judgements on the financial rating of organisations.

- *Employees:* an important part of any organisation is the employees. It is increasingly common for employees to have shares in the company they work for. An important part of management is employee relations, and staff should be kept appraised of the financial performance of the company. This has an impact on the people who work in the organisation and providing information helps to reduce uncertainty and increases morale and a sense of belonging.

Listed companies

The Financial Services Act 1986 provides the framework for regulation in the UK securities markets. Companies that are listed on the Stock Exchange have to abide by the regulatory framework set out in the Act. The Exchange regulates the flow of information from listed companies to ensure that all users of the market have simultaneous access to information. Under the Exchange's rules, companies have to make a variety of announcements to the market, particularly if the information could have an effect on its share price. To ensure that company information is available to all market users simultaneously, the Exchange provides a service for collecting, validating and publishing all company announcements: the Regulatory News Service (RNS). Companies on the Exchange's markets may submit announcements to the Exchange's Company Announcements office 24 hours a day – although announcements are only transmitted between 07:30 and 18:00 BST

The annual report and accounts

The annual report of a company is normally the key financial communication document of the year and needs careful consideration. What issues must be borne in mind when producing this publication? Prestige? Information? Public relations value? Such reports are vital communication tools and need to be directed towards the appropriate stakeholders. Their readership extends beyond the shareholders and analysts to other interested parties such as MPs, investors, employees, journalists, etc. The design of the report should take all these external parties into account. The annual report can become part of the overall public relations strategy of a company.

Design strategy

The design strategy for annual reports varies considerably. There are those that are elaborate and formal and some that are low key and casual. The design should have one objective in mind: to get the document read, with the rider that it is understood by those who read it. Not everyone can read or understand a balance sheet or the figures contained in annual reports, so there should be an attempt to provide information in a non-numerical format in addition to the standard official format. The aim should be simplicity and readability of language. The text elements should not be long-winded and the text size should be readable. It is important to remember that words can convey pictures and images. There needs to be a link between the words and the graphics. A failure to link the two often suggest that there is no link between the writers and graphics artists in the production of the report. The other major danger is to get carried away with the design of the report and to forget the communication aspect.

Some points to consider

- The cover sets the tone and draws the reader to the report. It is important to ensure that this is achieved.
- The opening pages tell the full story. If the reader does not go beyond the first few pages, the story should be as complete as possible. Bullet points should identify the key features of the company, and financial features should follow a similar format.

future.

- Senior management should be more than a group of pretty faces. The competence of senior management is a primary reason for investment. The biographies of the senior managers, detailing their experience and competence, are more important than their pictures.
- Feature sections should support the overall theme of the report. This section should demonstrate the competence of the company and how it will continue to exploit these opportunities.
- Financial information should be user-friendly while still within any statutory framework. Any variances, charges and exceptions should be clearly spelled out and referenced within the financial statements.
- Financial summaries should also be included as a means of showing progress. There should be an indication of how the company is doing in a five- or ten-year period.
- There needs to be an assessment of the overall market in which the company is operating, with forecasts of perceived market changes, both sectoral and geographical, and how the organisation intends to react to those threats and opportunities.

Investor relations

The shareholding population is increasing across Europe, largely as a result of the growth in privatisation. A new breed of shareholder is emerging that needs to be considered in a different light. The investor information required by this type of shareholder is a cross between a marketing exercise and public relations. One of the ways in which this is dealt with is to adopt a more customer-orientated approach towards shareholders. The investor receives dedicated communications about the activity of the company. This may take the form of an investment news-letter or similar correspondence. There is considerable opportunity for cross-selling by developing brand loyalty. In instances where there are consumer products there is scope for providing shareholding discounts on products. This increases selling opportunity as well as improving brand loyalty. A housebuilder could consider offering incentives to its shareholders or investors to purchase houses.

Investor relations is not only about selling but also about providing information on the company. It is about presenting the company's interests in the best possible light. The objective is to develop in investors a sense of loyalty towards the company.

Summary points

- Investor and financial relations is a specialist area
- The stakeholders involved in this particular area are more demanding and critical in their approach to the company. They expect communication to be prompt and unambiguous.
- There is more regulation of financial communication, particularly for listed companies.
- The use of the annual report as a communication tool should be carefully considered.
- The development of an investor relations strategy should be considered as a means of creating brand loyalty.

'The Powers That Be'

Introduction

This chapter explores the way in which construction companies can improve their interaction with government.

The construction industry like most other industries is influenced by the activities of government. The need for interaction with government is based on two principal questions: do the activities of government exert a detrimental influence on our activities and interests currently or in the future? Could government policy benefit us currently or in the future? The answers to these two questions are uncertain but we are aware that government impinges on our lives and organisations:

- Through legislation controlling our earnings (taxes)
- By controlling the environment
- By allocating government funds to priorities it decides upon
- By regulating the manufacture and distribution of products
- Through the regulation of business and employment
- By influencing the economic prospects through spending and regulation.

The impact of government is present in every part of our personal and organisational lives. The decisions that are made on the industry and organisations are made in an environment away from the confines of the organisations or industry.

Considerations

All individuals have a democratic right to approach the government if they so desire, but this right has to be exercised in a professional, cost-effective manner, with minimal disruption to the administration. The ideal situation is one in which individuals work towards achieving benefits for both the administration and the company.

There are four basic steps to consider when approaching government (Miller, 1990):

(1) Monitor
(2) Talk
(3) Argue
(4) Apply pressure.

This approach is designed to avert and resolve disputes between the government and its citizens. The idea of monitoring is to prevent the need to take drastic action as it allows for intelligence gathering and gives the government time to correct any misinformation or to change policy. By adopting this approach companies respond to government at their own pace rather than reacting to events. If the monitoring phase does not provide a satisfactory response, then the organisation moves on to the talking phase. This phase requires a well developed strategy and asks the questions *who, how* and *when*.

Who makes the decisions, who can influence decisions, and to whom should the case be made?

An important part of the strategy is to focus on who makes decisions and who can influence decisions. The power structure behind the issues of concern needs to be identified and the decision makers behind them targeted. Does the real power lie with the ministers, civil servants or policy units or with Parliament? The majority a government has in Parliament will influence the direction of the power. The important issue is to target the appropriate source of power.

How do you make the case, and how do you monitor its effectiveness?

The *how* is concerned with the manner in which you make your case. This is more than a few whispered words in the ears of someone influential. The research from the *who* should provide a better perspective on the manner in which decisions are taken and the mentalities that are prevalent. It does help if you have access to people with power, but this is not the overriding issue. The manner in which the case is developed and put together is far more important. It must be put together well, it must be targeted at the right people, and it must be put within an acceptable time frame before there is any hope of its being considered. The case must also

When do you make the case, monitor, lobby, argue or apply pressure?

Timing is often ignored, but in the life of politics and government it is a crucial factor. A case in point is when a Green Paper is published. Attempts to influence policy need to be made before the Green Paper is published, so effort should be made as early as possible. Effort after the Green Paper is published should be a continuation of an overall campaign and not some last-ditch attempt to influence policy.

Timing also means knowing when to undertake a particular activity. An understanding of the daily, weekly and annual cycles of government is required. It is important to know when recesses are, when parliamentarians are easily contacted, when ministers take holidays, and when Prime Minister's Question Time takes place, for example.

The approach to government should be established on a sound strategy that is based on good preparation, targeting the decision makers and getting the timing right. The remainder of the chapter looks at more techniques for successful government interaction.

Where do you obtain information?

Information gathering is a central theme of any form of engagement in government. It is crucial for any organisation engaging government to be aware of the policy, planning and political attitudes of government or, more importantly, how these are likely to change. The questions, committee investigations, statements and reports are a crucial part of any strategy aimed at Whitehall and Westminster. The level of information can help determine whether any potential conflicts are likely to emerge.

A fairly comprehensive approach to the gathering of information is needed. The idea is to keep informed about what is going on. Data gathering is an important issue but is also a costly exercise. The following represents the basic set of data a firm will need if it seeks to engage government.

- *Who does what*: officials, ministers, MPs, peers, committees associated with the organisation's area of interest.
- *What they are doing*: government commitments, policy changes, legis-

lative progress, investigations and reports, speeches, statements, Green and White Papers in areas of concern.

- What the motivation is what is the attitude of government and officials to the organisation's areas of interest? What is the level of priority and concern in areas in which the organisation is involved?

This requires a wide range of information and most of these sources are readily available.

Government publications

There are, for example, a number of sources that allow for careful targeting of ministers and officials. *The Civil Service Yearbook* (HMSO) is the most important guide as it lists all departments with their structure, and identifies the majority of key officials. Each department also has an information office which issues statements, briefings and speeches. The Central Office of Information (COI) produces details of all departmental information officers. A useful idea is to write to each departmental officer and get included on the mailing list for departmental information. Officials may also be contacted directly. Another source of departmental information can be obtained from the parliamentary branch of each department; the branch acts as a link between the department and Parliament.

The public face of the Government is Parliament. It also produces a large amount of information. *Dod's Parliamentary Companion* performs a similar function to the *Civil Service Yearbook,* in that it contains the names of all peers and MPs with biographies, photographs and other details such as constituency lists and members' interests. More detailed information is to be found in *Parliamentary Profiles* ('Roth') and in the *Register of Members' Interests.* In the past the latter contained omissions because disclosure was not mandatory, but after the publication of the Nolan Report it has become more comprehensive.

The business of Parliament is well documented, with each house producing an Order Paper, listing the day's agenda, and Hansard, an account of the proceedings in both houses. A Hansard for each standing committee is also produced. Both Order Papers and Hansard provide a wealth of information at a glance and obviate the need to spend time sitting in public galleries listening to debates. With practice it should take only a short time to review both documents although some of the longer debates may take longer.

Order Papers and Hansard are expensive to buy, so how to make best use of them? An efficient government observer should be aware of the statutory changes long before they are brought to Parliament and will not be concerned with a great deal of the information presented in Order

Register of Members' Interests and comments in the press. Hansard some-times provides useful material when responses to ministerial questions are made. Despite attempting to say as little as possible, answers may contain information on future publication dates of White or Green Papers and policy announcements. Information from debates of relevance to the organisation is useful in that it can help in the compilation of a list of MPs who support or oppose the organisation's position. Standing Committee Hansard is useful if there is a requirement to follow the progress of leg-islation in detail. In such a situation it is also useful to get a copy of amendments tabled for consideration and to cross-reference these to the government department involved to see if these are likely to be accepted by government.

The House of Commons also publishes a *Weekly Information Bulletin* which lists the forthcoming week's business in the Chamber and Select Committees. There are tables showing the progress of private and public bills, lists of the previous week's official publications and select com-mittee reports, government responses, individual transcripts of evidence and membership of standing committees when changes are made.

Another weekly publication, which is produced when the house is sitting, is the *House Magazine*. This is the in-house paper of Westminster and has an editorial board composed of MPs and peers. Each issue deals with a particular theme and contains articles from relevant ministers and opposition spokespersons, with views from backbenchers. It also lists parliamentary meetings and details of party spokespersons. The maga-zine is distributed to every MP and peer, which makes it an ideal target for corporate and pressure group advertising. This publication is open to public subscription.

Online information

The use of the information superhighway has provided additional access to information on government. Dedicated web sites for both houses are available as well as a large number of key ministries. In addition, the major parties have their own web sites which contain useful information on contacts and party publications.

Houses of Parliament Home Page
(URL http://www.parliament.uk/hophome.htm)

This site is maintained by the Communications Directorate, United Kingdom Parliament, Westminster. Here you will find information about the United Kingdom Parliament, the House of Commons and the House of Lords. The House of Commons site also contains information on Hansard reports of debates, public bills before the Lords and select committee reports.

UK Official Documents
(URL http://www.official-documents.co.uk/menu/uk.htm)

This service has been developed to assist users to locate official documents.

Her Majesty's Stationery Office (URL http://www.hmso.gov.uk/)

This gives information on Her Majesty's Stationery Office, Acts of Parliament, Measures of the General Synod of the Church of England, Statutory Instruments, the Copyright Unit, the Public Library Subsidy and Her Majesty's Stationery Office itself.

Other services
(URL http://www.butterworths.co.uk/content/links/51/130/index.htm)

This site contains excellent information about the law and other related internet services. It also carries information about the UK, EU and other related internet sites.

The use of the internet will make information more widely available in a cheaper and faster form. In using any electronic information service, the most important issue is not the information available but how often it is updated. Recency is a far more critical issue.

Other information sources

There are other published sources of information about government, legislation and public policy. The first is the press whose journalists are allowed to work in Parliament, and other specialist reporters who conduct interviews or undertake specific policy analysis. Editorial comments are less accurate as they tend to be predictions of changes; however, revelations about policy, ministerial intentions and backbench attitudes are more important. Many broadsheet newspapers list the work of both houses and standing committees on Mondays. Another source of information is any good law library which should contain a set of current statutes and statutory instruments along with any judicial decisions covering the legislation.

Another means of gathering information is through developing

not obliged to provide information unless there are bona fide reasons. They can help by recognising situations in which there may be mobilisation of opinion to act against the department. There is also the option of developing close relations with the main political parties as they are likely to be in tune with what is happening on a policy front. Most of the main parties have a headquarters which would research and collate information on Parliament. Adopting an aligned position may restrict access to other sources, particularly if there is change in power structures.

Advocacy

There needs to be a two-way flow of information between government and corporations. Governments need information to indicate that their policies are not formulated in isolation and based on misconceptions. Officials do not wish to jeopardise their positions by producing unrepresentative advice to ministers. MPs do not wish to risk their seats by endorsing policy or products that may lose them votes. Those who need to influence government must provide information that is accurate, appropriate and timely.

The information gathering process should help identify decision-makers and those in a position to influence them. It is only through experience that the correct officials, MPs and ministries can be targeted. There are a number of categories of targets:

(1) *Ministers* in the departments of greatest interest.
(2) *Ministerial assistants:* who are close to the Minister and can ensure that the information is conveyed to him or her. These may be parliamentary private secretaries or special advisers who help form ministerial opinion.
(3) *Officials:* these are divided between those who operate in a sponsoring role for a department and those who may oppose or object to proposed policy. Picking officials at the right level is important.
(4) *MPs:* at a constituency level, as members of select committees, as potential supporters or as opponents of the organisation's interests.
(5) *Peers:* those who have indicated particular interests that are similar to the issue or who are on committees.
(6) *External organisations:* these are influential organisations whose

representation and advice may help shape policy through forming opinion, through briefing, or through taking a role in advisory committees. This is a diverse group and may range from charities such as Age Concern; employers' organisations, for example the Confederation of British Industry (CBI); to trade unions; and pressure groups such as the World Wildlife Fund (WWF).

(7)   *Other targets:* there are other organisations (e.g. the CIB and the Royal Institution of Chartered Surveyors) that may require information about a major topic. They play a role in relationship building and in helping form opinions.

To which of these targets do you make representations? Appropriateness is an essential feature: too low a level and the representation will carry little weight; too high and access may be difficult and the representations will be passed down to those with sponsoring responsibility rather than to opinion-formers.

Why is there a need for government interaction?

The Government wants to be associated with success. Visits, interviews or other engagements with successful organisations allow it to explain policy and provide an opportunity for publicity for initiatives that the government is supporting. These contacts may range from private meetings, site visits, and opening projects to large set-piece events where information is exchanged and there is an opportunity for publicity. Such contact also allows the Government to gain feedback from interested parties. There must be an underlying reason for the contact: relationship building is simply not an option as the higher the level of contact the less likely the amount of time that is available.

Officials in departments also require information on how future external campaigning may influence the department. There is a need to know whether any future threat may emerge to the department and the minister through the galvanising of opinion. Policy is not constructed in isolation and officials need to take account of the widest range of views in formulation and to turn to external bodies for opinions. By doing so they discharge their duty of consultation. There is a wide variety of organisations that make representations to government; the net result is that officials are cautious and wary of representations made to them. It takes time for them to be satisfied with the credentials of external organisations who offer information. The lesson is to make representations to government long before any consultation period. Enough time must be allowed for the organisation to gain a reputation for providing valuable infor-

At a constituency level, MPs need to be aware of what is going on in their area. There is a need for interaction in order to keep the electorate happy and to ensure that they are aware of the changes that may be taking place in public opinion. Interaction and support for local industry provide a public means of making contact with constituents and of gaining publicity. The principle of being associated with the positive is prevalent in this interaction as well. Issues that are perceived to have a negative impact on a constituency are likely to be raised by the local MP. Many MPs sit on committees or act as spokespersons for the party and need external information to form opinion. Campaigning MPs need as much information as possible to take on departments or major interest groups without being out-debated on points of fact.

Why bother?

- Contact allows government to understand the objectives of the organisation.
- It gives decision makers guidance on the strategic objectives of outsiders.
- It acts as a means of ensuring that policy is kept on a particular path.
- Contact allows the organisation to forge friendly relationships and to defuse the impact of potential opponents.
- It offers the possibility of shaping opinions in advance of any policy change.

Contacting targets

Contact with government can be either personal or impersonal. Impersonal contact can take the form of letters, submissions, brochures, videos, company publications and organisational briefings. The personal contact is normally achieved through individual or mass meetings, visits, set-piece events and entertainment. Cold information provision tends to get the same treatment as 'junk mail'. There is a need to build up personal contact before presenting information. In certain cases, however, impersonal information is the only way to respond to situations such as responding to Green Papers. In such cases the information should be well presented and laid out. MPs have to respond far more to personal representations, particularly if they are from constituents or they are likely to be seen to be ignoring constituency opinion. A lot of the information gathering described in the previous part of this chapter has been about impersonal contact. There is now a need to secure access to convey the message properly.

Personal contact

At this point in the overall process the organisation will have decided who they need to inform through the research and monitoring process, and will have an understanding of those whose responsibilities and interests impact the organisation. The objectives of engaging government should also have been decided and the right targets identified with a properly timed contact.

Ministers

Ministers must be contacted directly but that contact can be reinforced by departmental advisors and by supportive MPs. The procedure for contacting a minister is the same for personal meetings or other contacts. The first step is to contact Ministerial Enquiries in the department and ask for the respective diary secretary. Explain who you are and whom you represent. State that you will be writing to the Minister for an appointment or extending an invitation to attend or address a meeting, etc. To avoid a clash of dates, you are trying to find out whether the Minister would be free on a certain date before you send a request. If there is a possibility that the date may be free, the diary secretary will advise you to put the request in writing to check availability. This allows the Minister to refuse any engagement that is considered to be not in the Government's best interests. The following points should be remembered when writing a letter to a Minister.

(1) Attention should be paid to the title of the Minister, particularly if they have received knighthoods or are peers.
(2) Show that you have gone through the correct procedure with the diary secretary and the date is cleared.
(3) Stress that the meeting is to be short as the Minister is unlikely to be available for more than a short while.
(4) Ensure that you indicate who you are and whom you represent.
(5) Indicate the theme of the likely meeting or event as it helps establish relevance. It is also helpful to indicate that contact has been made with officials relevant to the issue. If no prior contact has been made, then there will almost certainly be a referral to liaise with the appropriate officials. Meetings with ministers should not be the starting point but rather the culmination of activity. Do not lie about

Meetings with Ministers are not familiarisation exercises. They have to be specific and should have definite aims with a view to exchanging information. Influence on the Minister can be applied through the support of officials, the PPS (Parliamentary Private Secretary), or special advisers, or through the local MP. Liaison with other parties helps promote the chances of success. Support of officials can be extremely important as any meeting means more work for them in preparing meeting agendas, research, etc. Another means of securing meetings is for MPs to request meetings with ministers. They are more likely to accept a request if an MP agrees to attend the meeting.

If a request is for a visit to a project, it is in the local MP's interest to support the visit, as it will generate publicity for him or her. The MP will still be helpful even in opposition as the issue may be part of the overall interest area of the MP. A similar strategy can be followed with the Minister's PPS or special advisers as they are more likely to shape opinion than officials within a department. These techniques cannot guarantee success but they should help improve the chances of succeeding.

Special advisers and PPS

These should be approached in a similar manner to ministers on the basis that they are involved in areas that fall within the organisation's interest. The key idea is to stress that you will be briefing them on areas that are in the ministerial context. It is also useful to indicate that it is not just they who are being targeted but all levels of the legislative process.

Officials

Those who are in a senior position are less likely to agree to meetings unless the issues have been discussed with the respective specialist in their departments. The exceptions are to address public meetings, conferences or events of a similar nature. The usual manner of approach is through the written format. For the approach letter a similar procedure to that adopted for the Minister should be used.

MPs and peers

These are generally difficult to contact by telephone. Writing to them in the first instance is more likely to meet with success. Once again, the nature of the organisation, the purpose of the proposed meeting and the nature of the problem should be reinforced. It is important that the letter be kept concise (one page maximum) as the member is likely to receive a large quantity of mail. Meetings should be arranged in either London or the constituency. If you are inviting MPs to events in their own constituency, the invitation should come from the constituency branch of the firm and not some distant corporate headquarters. Try in such correspondence to be as personal or customised as possible, as mass-produced letters are often ignored.

Impersonal contact

In certain situations impersonal contact may be the only way to interact with government. This approach has more to do with not overcommitting the company and simply ensuring that government is well informed. Impersonal communication relates to letters, brochures, news sheets, reports, advertising and publicity material that is sent to government. As with all types of information, it must be relevant to the recipient. It is no use sending information to every department if the issue is specific to one area. Secondary contact is only sensible if a strong case for a particular issue can be made and its impact on the department established or clarified. Ministers not involved in particular departments have very little impact on the way in which policy is formulated in other departments unless it directly affects their own. Officials should be contacted following the appropriate protocol. Information should be sent in the first instance to the departmental specialists. Do not attempt to break the chain of command by passing information on to senior officials as this will inevitably end up back with the specialist.

Impersonal contact normally means that greater checking of the credentials of the organisation must be carried out. There must be information about the firm, why you are interested in the area, the relevance of the information, why you are contacting them and what you need. If these principles are not observed, information will tend to be treated as junk mail. It is also important to consider the amount of information the recipient receives. Lengthy documents that interest you may not have any impact on the legislator. The relevance of the document should be established. It is easier to ensure that any information that is sent is well received by the recipient by making contact with them prior to dispatch. There are guidelines that it is useful to remember when dispatching impersonal information:

- Clearly indicate what the information is about.
- Establish its potential relevance to the recipient.

Meetings

Like all other aspects of dealing with government, the preparation and timing of the meetings are important. What type of meeting is needed should also be analysed, for example a personal meeting, a visit or a set-piece event. The importance of the meeting to the prospective person in government should always be remembered. There are few further issues to remember.

- Do not invite ministers, officials or MPs to a private lunch or dinner unless you know them very well on a personal level. It is unlikely that there will be a response if this is not the case. If an invitation is made, then it is expected in the case of ministers and MPs that they are the principal guests and may be unwilling to share the billing with anyone else.
- The relevance of the meeting to the guest must be established. It is unlikely that the Minister of Health will respond to an invitation to open a prison. The size and level of importance of the organisation also influence the possibility of the visit. A minister is unlikely to visit an organisation unless it can help promote policy initiatives. It is more likely that a local MP will be interested in a local project than a minister.
- The timing of the meetings should also be considered. Develop a sense of understanding of the schedules of ministers, officials and MPs. The timing of the appointments to fit into the daily, weekly, monthly and annual calendars should be considered. For example, MPs are more likely to visit their constituencies on a Friday or Saturday when the House is in recess.
- It is important to remember that it is not possible to make appointments at short notice. Ministers, MPs and officials should be approached at least a month in advance of a potential meeting. It is always important to remember that you are competing for limited time in all cases.

The distance the target has to travel from their base will also have a bearing on the outcome of the visit. It is important not to forget that there are costs associated with visits or meetings. The main issue is that the company must be seen to get value from what is spent. As a rule entertain sparingly, concentrating on those individuals who have genuine power but who are not necessarily well known names.

Meetings are likely to be short and require you to get as much information across to the recipient as possible. Adopt the principle of stating who you are and what your interest is, explain the issue at stake, and say what is necessary. As a rule do not introduce any new material to the discussion as this will require further explanation. In most cases the briefing note that has been sent will form the basis of the discussion. Do not quote reams and reams of statistics as they serve little purpose in a focused meeting. Expect questions and try to generate a discussion. Try to remember in all discussion that these individuals or groups can influence the future direction and success of the organisation.

Do not be emotional in any presentation of material. Make your case as dispassionately as possible. It may be useful to minute as much of the meeting as possible. In the case of ministerial contact there will be officials present who will do so on behalf of the Government. The key strategy of any meeting of this type is to make sure the major points you wish to make are remembered. A useful target is three or four key points. Once the meeting is over, do not forget to send a letter of thanks reiterating issues that have been raised and promises made either on the part of the Minister, the MP or an official; and yourself.

Arranging a visit

Once agreement has been reached for a visit, a series of steps should be followed.

(1) Indicate who is likely to be present at the meeting, visit, conference.
(2) Determine the precise timing of the activity in which the individual has to take part during the visit.
(3) State who will meet them on arrival.
(4) Identify the person to whom they will be introduced.
(5) Find out who else is participating in the event; whether there will be other speeches; what is the likely content.
(6) A briefing note should be sent to indicate the purpose of the event and its key focus as items may be included in any speech that is being prepared.
(7) Check if the visitor has any special requirements.

It is also useful to remember that a high-profile visitor may attract attention from more interest groups than the media. Given the number of high-profile protests that have taken place it is useful to consider security

engagements is achieved. If there is no obvious agenda, there is little purpose in arranging meetings or events.

Promoting or amending legislation

Interacting with government is a means to changing legislation and the impact it has on the organisation. Interaction takes place at all levels within the legislative process. There should be an opportunity to influence legislation at every point within this process. There are three main routes to influencing legislation:

(1) *Party and ministerial pressure:* political commitments made by parties in their manifesto or as part of the broader political agenda; and ministerial commitment to ensure that these are carried out.
(2) *Pressure from officials:* who are able to convince ministers on technical and administrative issues. Officials have a great deal of influence on the framing of the legislation and on secondary legislation. They advise on interpretation and inadequacies in the process.
(3) *Pressure for MPs and peers:* this is achieved through private members' bills or by tabling amendments to legislation. If there is sufficient support in Parliament, then the executive arm of government may consider changing the legislative programme. The issue may be raised in the House of Lords in a similar manner.

The ability to influence legislation is clear but there is a definite need to engage the respective influences.

Common mistakes

- *Friends in high places:* this is the most common misconception. Those who think they can get their way because they know officials, ministers or MPs are mistaken. A sound case and an understanding of the process are more important before use of any contacts.
- *Entertain your way out of trouble:* this is often seen as the standard approach to solving problems or influencing people. A well researched case, conveyed in an efficient manner, works wonders and is more cost-effective than largess.
- *Act now, think later:* 'Research, research, research' should be the motto

when dealing with government. If the organisation does not know how the Government is going to react to a particular approach, then inadequate preparation has been done.

- *Parliamentary focus:* Parliament is not the be-all and end-all of government. Do not separate the elements of government. The structure of government involves all elements and an understanding of all elements is essential (Miller, 1990).

Outside help

Dealing with government is a complex issue. In order to keep abreast of the changes taking place in legislation there may be a need for dedicated resourcing. For many companies there is the alternative of using external sources. These include representative organisations such as the Institution of Civil Engineers (ICE), the Chartered Institute of Building (CIOB), the Confederation of British Industry (CBI), and the Construction Industry Council (CIC). Governments prefer to deal with organisations that represent an industry rather than with individual companies, and representative organisations tend to have easier access to parliamentary interest groups than individual firms. The collective experience and resources can be used to a company's advantage. There are also the drawbacks of not representing individual interests and having inadequate resourcing. If a company wishes to use this route, then there is a need for it to become more involved in the representative organisation and to make sure that issues of concern are raised.

There is also the alternative of engaging external government affairs consultants. These companies in many cases provide a lobbying service and focus a great deal of their attention on Parliament. They are not cheap to engage and may not be specialists on the industry of which the firm is part. There is no vetting system and there remains a question over the standards of service provided. Consultants that have a working knowledge of both Westminster and Whitehall provide a better knowledge on the workings of government. The best approach to appointing such advisers is to treat them in the same manner as other professional consultants and require a similar level of performance.

The use of external sources should provide value for money.

Communicating in Europe

The European Community (EC) has an influence on both individual and corporate lives in the UK. The EC is one of the biggest trading and

competition, employment rights, foreign aid, etc. Its impact is far-reaching and there is considerable scope for engagement.

Most people in the UK gain their knowledge of the workings of the EC through the mainstream media. Knowledge of the EC concentrates on issues that are raised in the media. Issues such as the Common Agricultural Policy, the introduction of the Euro, fish quotas and beef bans tend to be the areas that are most in the news. A great deal of the reporting on the EC raises fears over the impact of integration although much of our daily lives are not influenced by the EC.

On a more positive note, the EC has been responsible for changing many aspects of our lives. The EC has forced greater environmental control in areas such as water standards and cleaner beaches. It has introduced the Social Chapter which provides greater rights to workers and promotes equality in employment as well as greater competition and openness in many areas including construction. EC procurement practices have created more opportunities for involvement in Europe. Opportunities have been created through regeneration and redevelopment funds. The influence of the EC will continue to grow, and companies with international and global ambitions realise that their views need to be heard in this forum.

Understanding the power structure

The Council of Ministers

The Council of Ministers is the major decision-making body in the EC. Every member state holds the Presidency of the Community for six months in rotation. There are meetings of heads of government and senior ministers on a regular basis. Operating in parallel are ministerial councils in areas such as finance, foreign policy, environment, agriculture, trade, energy, tourism, etc. The meetings of these ministers take place on a more regular basis and key decisions are made by ministers acting almost as a cabinet for the community.

In order for any issue to be raised at this level it has first to be raised nationally. The Minister in the host country has to be engaged and the issue made part of the national agenda before it is likely to be raised at a European level. Thus lobbying, issue management and other forms of corporate communication must take place nationally in the first instance. Engagement on a purely European level should only be attempted under very special circumstances.

The Commission and the Commissioners

The Commission is the Civil Service of the EC and it is charged with the day-to-day running of the Community. It is responsible for drafting proposals for consideration by the Council and the Parliament. The Commission is charged with implementing decisions once agreement has been reached. It has more power than the British service to initiate and formulate policy without direct guidance from politicians. The Commission has powers delegated to it under the various treaties and by the Council, and can make decisions without further reference to anyone. It can, for example, fine companies or take governments to the EC Court of Justice for infringements such as breaches of competition rules.

A more important power of the Commission is its responsibility for proposing regulations and directives, which are the basis of community law. These are referred to the Council of Ministers and the European Parliament. Consultation can also take place within the House of Commons and the Lords. Proposals are sent to the Council, indirectly on to each national government and then to the European Parliament. There is an extensive consultation process and the Council may suggest changes and amendments or even reject proposals. Once the consultations are completed and a majority voted in the Community Parliament, the proposal becomes Community policy. The policy may take the form of a regulation, which is enforceable throughout the EC, or a directive, which requires national governments to pass implementing legislation.

To what extent is engagement possible? Organisations can only have an impact on the process when they are aware of the existence of draft proposals. Coverage of these proposals in the British media is generally poor. A source of information may be when proposals are presented to Parliament. European Commission offices also provide a source of information. The growth of the internet has made information about the working of the EC easier to access. The official EU web site is http://www.europa.eu.int. It contains a wide variety of information about the business of the EC and on all the major institutions and committees, and on policy and laws. There is an on-line service on public sector procurement. The information is structured into a series of databases that are easy to access. This service allows the user to extract information easily and in certain instances provides contact names of relevant EC officials. The internet site of the EC is well organised and is an extremely good source of information.

Co-ordinated action across the EC is fairly limited. There are certain groups or companies with European-wide interests that co-ordinate their actions across the Community. The trade union movement through the ETUC (European Trade Union Confederation) works together to promote

tions with cross-European stakes need to consider to what extent they need to engage the Commission. Many multinationals from other industries have EC communication and lobbying strategies as they realise the growing influence of the Community.

The Commissioners in a sense act as the 'cabinet ministers' of the Commission. They head the sections where policy is developed and answer any questions in the European Parliament relating to the particular policy. There is no reason why companies or pressure groups cannot let their views be directly known to the Commissioners. The principles of importance, relevance and European interest should be followed. Extreme care should be taken when approaching the Commissioners in this way: it is a privilege and should not be abused. It is better if an approach is made in conjunction with a Member of the European Parliament (MEP). The key point to remember is that in the EC you are dealing not only with particular political views but also with the different national interests.

The strategy of dealing with the Commission and the Commissioners is to make sure that issues are first placed on the national agenda. This will force the national government to make it an issue. Secondly, try to make sure it has a European dimension as this is more likely to be noticed by MEPs. Thirdly, try to develop a European alliance, which means that it is more likely to get on to the EC agenda.

The European Parliament

The Parliament is directly elected from each member country. Although no one group has an overall majority, there are large groupings such as the Socialists and the Democrats. It does not possess the power of the Council of Ministers or the Commission. The role of the Parliament has largely been consultative and advisory. It can debate all matters relating to the work of the Community and foreign relations with the Community. The powers of the Parliament have increased to be able to delay legislation or to force compromises between members of the Council of Ministers.

The Member of the European Parliament (MEP) operates on the same principles as a Westminster Member of Parliament. The issues of engagement follow similar lines. Information about the Parliament is important, as with the House of Commons. There are a number of sources of information available. For example, there is the Hansard of the Parliament, referred to as the *Supplement of the Official Journal*. There is a delay in the production of this document because it has to be translated into the official language. A record of decisions reached, called *Texts Adopted by the European Parliament*, comes out soon. On line electronic information about the European Parliament is easily reached and provides a quick source of information.

It is useful to remember that the average European constituency is approximately eight times the size of a Westminster constituency. MEPs also spend a great deal of time travelling between the Parliament and their own areas. Contacting them at short notice is difficult. In order to gain access to an MEP it is best to write to them locally, following the principles laid down earlier in this chapter. Like all good politicians they like to be shown in a good light and probably welcome local publicity.

Strasbourg and Brussels

Meetings at Strasbourg and Brussels involve a great deal of cost and effort. A more cost-effective way of getting an issue on to the European agenda is to raise the issue in Britain. Taking issues to Strasbourg and Brussels should mean that you are planning an international campaign. There, all the MEPs are in the same place and it is possible to gain access to members from other countries. Unsolicited approaches are likely to fail. A local MEP can pave the way to gain access to other members and to the Parliament. It is useful to remember that MEPs are not based at the European Parliament and may spend only a short time there. Last-minute approaches to shift agendas or change policies are unlikely to work as the detail tends to be resolved at committee stage. In order to have issues raised or to influence decisions, approaches have to be made before committee stage. MEPs can also pave the way for meetings with commissioners and other officials in Brussels.

Some points to remember

- Start any campaign/communication on the home front. If it is part of the national agenda, the Government may see it as worthy of discussion on the European front.
- Approaching the European Parliament or the Commission directly should only take place when a Europe-wide issue exists.
- Build multinational coalitions as they are more likely to get noticed.
- Despite the lack of attention in the popular media, the EC has a good information service on the internet and from its regional offices.
- An MEP can provide the same degree of help as an MP.

Summary points

- Understand how decisions are reached throughout the governing process.
- Parliamentary interaction is not sufficient: the executive also needs approaching.
- Research and preparation are fundamental to interacting with government.
- Information is essential to be successful.
- Establish the relevance and value of any interaction.
- Build relations and establish proper credentials: this is a long, time-consuming process.
- Lobbying government is the art of the possible; going with the flow is easier than being antagonistic.

Chapter 8

Communicating Community Involvement – What Will the Neighbours Say?

Introduction

Society is changing and looking at new ways of examining political, organisational and business behaviour. The socio-political lobby is becoming more influential and sophisticated. Community politics are becoming increasingly influential in shaping local and national decision making. Consequently greater accountability and community involvement are demanded by society. Business and development activity have the power to shape and alter the way society evolves through its actions and the direction of resources. It has a far-reaching influence on social order, hence the greater scrutiny of business by society.

The construction industry is not different in its impact on society. It has the same power to shape and alter the way communities develop. Unlike many other industries, however, it has the ability to exert a more profound impact through its end-products. The products of other industries are consumed and often do not have a long-term presence but the impact of the construction industry can be seen all around us.

The construction industry interacts much more with the society with which it comes into contact. The consequences of not taking due consideration of this block of stakeholders can be extremely costly and will have a negative impact on the success of projects. This chapter sets out some of the issues facing public relations in a community context.

Traditional views of community relations

The social compact between business and the community has a fairly established history. The tradition of benevolence and social care in the UK from industry dates back to the activities of Cadbury and Rowntree which tended to be steeped in Quaker philosophy. The tradition was that business and industry had a role to play in the social welfare of their employees and the communities in which they were founded. Social business is business'. This philosophy is based on the work of the economist Friedman. It follows the view that business should concentrate on what it is good at and not become involved in any social engineering. Its followers hold that the issues and problems that society face are not a concern of business and should be dealt with by the state and society. These views still have strong support from many who follow a market-driven philosophy.

Most businesses tend to fall into one of these two models. Donations to charity are the most clearly visible activity in modern corporate philanthropy. Private-sector funding tends to be about 'favourite causes', rather than activities associated with their line of business. This is a minimalist approach that devolves the needs for 'corporate citizenship' to charities and associated organisations. In an age in which there is a growing demand for industry to do more, are these existing strategies adequate? This also poses the question about the value of such activity. Contributions to charity are a good philanthropic idea, but how does this contribute to the business success of the organisation? How does this allow the company to engage its stakeholders in a more profound manner?

A new paradigm that can be used to improve corporate communications

The latter half of the 1980s and early 1990s saw the larger part of the developed western economies go through a recessionary period. Downsizing, retrenchment and restructuring became the principal activities in the drive to reduce costs, improve efficiency and become more competitive. In an environment that was associated with cost cutting and laying off workers, it was difficult to justify community involvement or corporate giving. This would be seen as hypocrisy by the many communities that had suffered as a result of corporate action. A new approach was needed that tied corporate philanthropy to a strategy which Smith (1994) describes as 'New Corporate Philanthropy'.

The new paradigm encourages companies to tie their community strategy and philanthropy to their business strategies. The business and welfare units join forces to develop corporate strategies that will give the organisation a competitive edge. This approach requires the company to back its strategic decision-making with action that supports its stakeholders. A simple illustration of this would be an organisation undertaking a development project that deprives a community of green space.

The company would then provide remediation for this by creating another space or facility, although it is not obliged to do so. This ties the philanthropic side of the organisation to its business activity.

This new paradigm asks organisations to adopt a long-term approach to community. The model generates an array of opportunities that are self-serving to the business and are a basis from which to generate public relations activity that supports the organisation's needs.

The need for a focus – the *Exxon-Valdez* case

The *Exxon-Valdez* oil-spill case highlighted the need for organisations to develop a focus in their activity. The Exxon Foundation was particularly well known for its educational activities and for being distant from the activities of its parent company. Indeed, the social profile of the Foundation was considerably better than that of its parent. Its activities were directed at building relationships with environmental groups and its consumers. Flaws in this philanthropic philosophy were not exposed until the *Exxon-Valdez* spill in 1989. Exxon adopted a reactive strategy after the spill as it had nowhere to turn to for advice on handling the crisis. This allowed Exxon to become an easy target for the anger of environmental groups. In effect Exxon had no relationships with environmental groups because the Foundation insulated the company from feedback from important interest groups. The company had no test bed of support in the key area of its business activity (Smith, 1994).

The adoption of a model that tied Exxon's core business activity to its philanthropic ideas and community relations would have given the organisation access to important stakeholders. Exxon would have been better placed to respond to the problem.

A parallel can be drawn with those construction companies involved in environmentally sensitive projects. Companies that have their own environmental policies are better placed to understand the sensitivity of such projects. In developing their own systems they are forced to understand environmental issues. This provides important intelligence to cope with projects that may have such problems.

Construction and the community

The construction industry has often faced hostility over its activity. The positive aspects of construction activity are rarely seen, but the negative aspects are accentuated. The 'blots' and 'carbuncles' gain the greatest publicity.

interest groups and communities. These confrontations are by no means isolated and extend to other types of construction and development projects. The coalitions that oppose construction have also gained additional support from parts of the public that feel that their quality of life is threatened or compromised. A new, more affluent and influential group of people are joining elements of the anti-construction lobby in certain cases.

The industry also suffers a poor image by being seen to be dirty, noisy and disruptive. It also suffers the effects of 'short termism': here today but gone tomorrow without concern about the aftermath. With these credentials, construction activity is often not welcomed by the communities in which it takes place yet the industry is inextricably linked to the communities in which it operates, through the planning processes, labour, resources, production processes and the final finished products. As communities become more sophisticated, they acquire the means to disrupt, delay or ultimately stop projects from achieving completion, or conversely to help expedite them. It would therefore be logical to acquire the support of local communities to ensure projects are completed satisfactorily. In order to function more effectively, the industry needs to consider different strategies when interacting with communities.

Being considerate

Clients with a social remit require greater social input from their contractors and consultants. Local authorities are at the forefront of this drive. They are looking for planning gain before projects are allowed to proceed. Indeed, the Considerate Constructors Scheme was established largely as a result of a local authority initiative in central London. Large retailers, particularly the supermarket chains, are concerned about the impact their construction projects have on their local communities. They are present in locations in the long term and do not want their projects to create any disharmony or negative publicity within the communities in which they operate.

A structured approach to social policy is required in order to meet this need and provide viable deliverables. A more focused community strategy to develop closer relationships with target communities would therefore seem ideal to help in the implementation of construction activity. Can a new, focused approach that takes cognizance of the concern of people be more successful than the current approaches that often ignore the concerns of society in general?

What are community policies?

The development of a community policy is concerned with the orientation of the company to key external environment influences. *It is also a reflection of the norms and values that an organisation adopts.* An emerging theme is that world class companies develop community policies within the localities in which they operate (Kanter, 1995). Corporate involvement in the community is not only geographically and project specific, but involves a philosophy that requires a long-term outlook. This is contrary to the short-term approach of construction projects and would suggest that the strategy needs to operate at two levels: at the project-specific level, but also at an industry or wider environmental level. (See Fig. 8.1.)

Fig. 8.1   Community policy framework in construction (Moodley and Preece, 1996).

The industry/environmental strategy

This strategy is aimed at issues that influence the company irrespective of the project or location. Community policy aims to support activities that are of strategic interest to the organisation, as well as its culture and the values of its leaders. The other influential factors are general social, economic, political and environmental issues. This part of the emerging strategy is concerned with the long-term approach needed to make community policy meaningful.

This part of the strategy relates more directly to the corporate ethos of

features of 'excellent' companies. A company committed to the education and training of its workforce would be able to translate these activities into a community policy. A literacy initiative, a skills development programme, etc. may form the basis of a community strategy. The new approach meets the needs of corporate goals and also actively engages an important part of the community.

Project strategy

The project strategy is aimed at local and project concerns. Ideally local policies fit into the industry-wide strategy but sufficient flexibility has to be built in to allow implementation at a project level to be meaningful. Local economics, values, politics and infrastructure play an important part in community strategy at project level.

The project level strategy in construction is important due to the contribution that project success makes to the overall success of the organisation. Projects are the production mechanisms of construction companies. Community interaction is also greatest at the project level as the organisation is often operating at the heart of a community. The purpose of developing a strategy is to take cognizance of many of the issues that communities feel need to be addressed when construction activity enters their environment. Issues such as pollution, noise, disruption, safety and privacy are often raised as problems that communities encounter. Standard procedures conceived in advance of execution of projects will ensure a smoother implementation. Relationship building at project level can also ensure that disruptions are kept to a minimum.

Making a business case for community involvement

There is always a difficulty in finding resources and making a case for activity that is not directly production related. One has to ask the question: is there a measurable business case for corporate and project community involvement in a 'bottom line' strategy?

Improving corporate image

Positive community policies can improve a company's market position at both corporate and project level. The immediate impact of a community policy is on the profile and image of the organisation. Involvement in community policy increases the profile and recognition of the company.

Recognition creates a brand identity associated with the particular activity. The image of the company is enhanced through the work it does. At a local level the image of the organisation is directly associated with the projects they undertake. The manner in which it operates at this level will have either a positive or negative influence on the organisation. The identity is still enhanced irrespective of the message that is generated.

Access to a wider network

Involvement brings the company into a wider network of interaction. The power of community politics, particularly in the planning and approval process, should not be underestimated. Local people power can have a much greater influence on the direction a project takes. The development of technology means that access to information and the transmission of information are easier. A developed social policy should provide an additional resource to test ideas and concepts before entering the formal system. The American oil company Arco have successfully used this method by working with environmental groups. The primary benefit of such associations is the feedback from groups directly associated with the industry (Smith, 1994).

A competitive advantage

Competitive advantage can be gained where clients require construction projects to provide more than the facilities. Organisations with a community policy are better able to market themselves to such clients. Community involvement indicates that organisations are better able to operate in partnership arrangements with communities. These organisations enter into partnership arrangements with a clearer vision of what is achievable under such arrangements and with a more realistic approach to the expenditure involved in these processes. This is evident in new trends in regeneration projects. This competitive edge is further enhanced by the organisations' improved image and access to existing community networks.

More attractive to recruits

Employees, particularly young ones, are increasingly concerned about the public reputation of their employers. There is a stigma attached to organisations with negative credentials, particularly among entry-level graduates. Companies that have sound community relations are more likely to attract able graduates. Involvement in activities outside the organisation acts as a mechanism to boost morale within the organisation. The growth of the 'united way' and similar schemes shows how the

for training managers. Ultimately the people within the organisation feel good about the organisation they work for.

More ethical investment

The general public and ethical investors are looking increasingly at the activity and behaviour of organisations. People prefer to avoid companies that are regarded as not socially responsible. The growth in ethical investment funds also indicates that ethical investors are looking more closely at the activities in which organisations are involved. Involvement in community projects increases the ethical dimension and social responsibility profile of the organisation. As prices, quality and service differences narrow, organisations will have to find new ways of differentiating themselves. In the marketplace, the public's perception of an organisation is critical to its future viability. It is interesting that the UK supermarkets are now selling organic produce and are sourcing their products under 'fair trade' schemes. As an industry sector they are probably closer to consumer opinion. Large supermarkets are also particularly concerned about the image that is created when they construct new facilities. They are therefore keen to minimise disruption to communities and to avoid presenting a bad reflective image from the construction companies they employ.

Implementing the community policy – setting goals

The starting point for the community policy is vested in the corporate goals and values of the company. The strategic objectives of the company should include a clear altruistic commitment. All subsequent action can then be evaluated against these aims. Clear strategic leadership is also necessary, as has been the case with both John Laing plc and the Costain group. The CEOs of both organisations were instrumental in driving the need for more comprehensive community policies (John Laing Group, 1986–1996; Costain Group, 1994a).

Analysing the opportunities

Once the commitment to a community policy has been established, the next stage is to decide on the possible courses of action open to the organisation. The community activities that are adopted will provide the greatest benefit to the community. The current social issues and needs are the base from which the policy can proceed. The aim is to support a community issue that would be beneficial to the organisation from more than a purely publicity context. These activities would contribute to the business interests of the organisation. A generic corporate approach has to be thought out to give the organisation the results it needs.

Project-level strategy should be easier to develop as many of the issues that need addressing at a project level are well known. Much of the project level activity is governed by legislation but there is scope to ensure corporate values are also included. A company may have a general policy of promoting education and training. At a project level this can be achieved by incorporating site safety into an educational package. This could then be taken into local schools not only to act as safety training but also to ensure that the organisation is known in the community. The profile of the organisation is raised while making sure that safety is improved. Issues related to community safety communications are examined in further detail in Chapter 11.

Commitment of resources

An essential part of the community strategy is the commitment of resources to the policy. This may take the form of money, staff time or equipment. The commitment of resources for what are often seen as intangible returns is often the most difficult decision to make. Inadequate resourcing and the lack of genuine commitment to the policy will lead to failure of the scheme. If the policy is tied into the corporate goals, then the schemes are more likely to have adequate resources allocated to them.

Gaining internal support

The construction industry is noted for its instability and resistance to change. In a climate of restructuring, downsizing and redundancies, it is difficult to convince employees that a community policy is important or that the organisation is concerned about the people around their projects. The case for the policy has to be sold if there is to be commitment to its implementation. The managers and staff that are required to implement these activities need to be aware of the importance of their actions. They need to know too that there is organisation commitment to the policy and that it will not be axed during the next cost-cutting exercise. Selling the idea is not enough: there must be back-up support and resources. Training and educating those at the cutting edge of

in place. Instinctively, most project personnel are more directed towards hard targets, but if the case is sold convincingly then there is likely to be more positive support.

Gaining external support

Once the policy has been sold internally it has to be tested in the environment in which it will operate. Involvement in costly schemes that do not satisfy community needs is futile and often counter-productive to the aims of the policy ideals. Consultation with interest groups on the actions and activities to be undertaken is important. The practicalities of the implementation, the problems and the possible out-comes are more likely to be identified during this process. The principal danger is that there are often a wide variety of possible interest groups that may be operating to their own agendas, therefore care must be taken to ensure the programme is not hijacked. A useful dialogue will, however, ensure that the policy can be refined before being introduced. In building up the dialogue, remember that it must fit into the interests of the organisation.

Operational issues

The policy should be fairly clear at this stage. Its aims and mechanisms should have been identified after the internal and external consultation stages. It is unlikely within the construction industry context that these activities will be executed by people exclusively involved in community policy. Construction project managers and regional managers are more likely to be the group that is involved in many of the day-to-day decisions on these activities. These individuals will have to be made aware of the objectives of the community policy and how they help in developing the profile of the company.

The business case for the community policy must be made. A clear and easy-to-use management interface has to be developed between those responsible for the community policy and those involved directly in its implementation. The support system for those staff who are not familiar with the policy should be introduced with a view to assisting in decision making and action plans. Visible signs of resources must be evident if people are to believe that such effort is viable.

Community policies in construction

The foundations of community policy in construction have traditionally concentrated on four main areas:

(1) The environment
(2) Regeneration
(3) Enterprise and employment
(4) Education and training.

Construction companies are often involved in many of these activities so there is an ideal springboard from which to develop a fully fledged community policy.

Environment

An environmental policy is almost a prerequisite of every major construction company. This is the biggest battleground between the construction industry and the public. A company that is likely to be involved in activity that is environmentally sensitive should focus on community schemes that will bring them into closer contact with interest groups. This will reinforce the commitment of the company to the environment and will provide a useful mechanism for obtaining feedback and advice on how to deal with future environmental problems. A proactive response to environmental problems is a good way of setting the tone for positive community relations.

The investment in these activities should be targeted towards meeting the corporate goals of the organisations. Adams *et al.* (1991) identify the areas on which to focus when developing an environmental policy, namely: renewable resources, pollution, energy consumption, fossil fuels, transport and pro-environmental initiatives. It should not be unthinkable for construction organisations to work with such bodies as the Royal Society for the Protection of Birds (RSPB) on species protection. This will improve the environmental profile of the organisation. Furthermore it will give the company access to the extremely valuable experience of the RSPB, which is involved in preservation and rehabilitation work. This association could become extremely useful when the organisation is faced with environmental concerns.

Another useful way of showing how serious a company is about the environment is the production of an annual environmental report with an environmental balance sheet. This document could stand alongside the organisation's annual report and be a clear indication of the environmental credentials of the organisation. This will of course draw much

The environmental part of the scheme can be introduced at a local project level as well. Issues such as noise, dust, litter and graffiti should become part of the environmental response to any project. The environmental appraisal of construction projects should also look at such measures as landscape rehabilitation, removal of waste and decontamination. This is usually up to the client and is normally built into the contract as an element of the works. A first step in the right direction would be signing up to the 'Considerate Constructors Scheme'. The scheme requires contractors to:

(1) be considerate to neighbours and the public;
(2) protect the environment;
(3) be clean;
(4) be good neighbours;
(5) keep sites tidy;
(6) stay safe;
(7) be responsible;
(8) be accountable.

While this scheme does not fulfil every virtue of project level community policy, it does make a useful starting point. A community commitment to improving the environment is created that is closely linked to the key business objectives of the organisation.

Regeneration

Construction is a key driver for regeneration, particularly in urban areas. This offers another opportunity to develop a community policy and link it to the aims of the organisation. The traditional approach to regeneration was to develop the infrastructure and then allow the communities to fit into the schemes. This approach was identified as being flawed, and the goal of new regeneration schemes is to set up a partnership between the local community, the developers, the constructors and the government/funding agencies.

Regeneration schemes provide construction organisations with a unique opportunity to build a long-term presence in the community. Opportunities exist to develop the partnership between the community and the regenerators during the planning, design and construction of the work. This is easier for organisations with an active involvement and a sense of ownership of the scheme.

Employer

The construction industry as a major employer within communities can also be brought into the remit of community policy development. This relates particularly to the amount of local labour that is used allied to any additional training provided. Improved skills make people more employable and are a value-added benefit to society. The number of new jobs created and of locals undertaking these jobs are good indicators of commitment to community economic regeneration. Statistics on local people involved in the project can be used in publicity material about it.

Enterprise, training and education

General education and training need not be construction specific to help society. Community involvement means attention to areas outside construction. For example, companies may decide to help develop literacy programmes, set up child-care facilities or provide sports equipment for the local community. The Costain group has developed a scheme called the Costain and Young People Joint Venture (Costain Group, 1994b). The programme aims to involve the company in education in the community. It is driven nationally by the organisation and implemented where the company has work. The scheme centres around the work the company does and is the basis for the learning process.

It is also useful to provide information to local schools on the construction project. Relations can be built by inviting the school on site visits. The issue of site safety and the dangers of construction sites can be a further educational tool. It is also possible to organise design competitions that increase interest in the projects. The greater the interest in the project, the less likely the incidence of problems on site.

There is a need to develop a sense of ownership of the site. An example of this is the work of a large civil engineering contractor who teamed up with the staff of a local school to exploit the positive opportunities it could offer the pupils during their road improvement scheme. The project was going to create a lot of dust and disruption outside the school but the partnership between the teachers and engineers meant that pupils would benefit from the experience. They gained a first-hand understanding of everything from construction to materials. Pupils were able to tour the site regularly and took part in competitions about the contract. The contractors also supported the school with donations to a special unit.

Another useful tool is the provision of company skills to benefit the community; for example, legal aid clinics provided by company lawyers, or help with promotion by the marketing department to local businesses. The Laing Group uses this philosophy as part of its management training

While these have been the traditional grounds for construction development, there are potentially other areas that could be of interest to the organisation. The aim of the process is to identify areas that relate to the organisation's overall business aims.

Customer service/community service orientation

A tried and tested approach to community interaction is to develop a customer service orientation for each project. Information about the project should be made available to interested parties. The site information boards should clearly state the nature of the project and its probable duration. This gives the community an explanation as to why and for how long their lives are going to be disturbed and tells them what benefit they will get from the project. This information could also be made available at central locations such as libraries and shopping centres. Local residents likely to come into contact with the project should also be informed where possible.

As part of developing a service orientation ethos there should be a complaints or information hotline for the project. A project 'liaison engineer' should be appointed for larger projects. If the cost of having a dedicated line is too expensive, then a corporate level decision about the importance of such a service should be investigated. While some in the construction industry may balk at the cost, it is imperative to remember that most customer-orientated industries have a dedicated information and complaints service. Setting up such a service is no good if there is no follow-up to complaints. It is also the sign of a mature and confident company that a contact person is provided or allowed for on the project.

Before the start of a project where community-sensitive works are involved (i.e. urban roads), the client, contractor and other parties should establish this 'liaison forum'. It is not just the responsibility of the contractor.

In developing a community orientation the aim must be to minimise the impact of construction on the local environment. Thus the organisation should attempt to reduce any damage to the immediate environment such as air pollution and noise. Paying attention to these will also reduce the number of complaints to the organisation and to the local authority involved. The hoardings on the site should be kept in a tidy condition, free from graffiti and unsightly posters. It is also useful to remember that not everyone wants to see large parts of a messy site. If you are going to provide viewing, the use of viewing portholes or windows is a good idea.

Promises must be kept as a demonstration of good faith. When the company promises not to remove a tree it should keep to its promise. If a promise is made to create a playground, that commitment should be fulfilled. Nothing sours community relations so much as bad faith

Disruption to communities also occurs outside urban or built-up areas. On road projects the dreaded cone has almost become a feature of the landscape. The use of the local media to inform motorists of diversions, delays or other distractions is also part of the development of customer orientation. Information should be provided on a continuous basis. Most areas in the UK have radio stations that feature travel updates during the mornings and evenings at least. Using these sources helps reduce the impact on the community and is largely the client's responsibility. The smaller the degree of disruption of local life the more successful the project is.

Direct action/lobbying

Construction organisations are the visible face of construction. The public are often unaware that these organisations are mainly the executors on behalf of clients. This often brings the public into conflict with the organisation. Action against the organisation manifests itself at two levels, the project level and the corporate level.

At the project level, opposition, protests and direct action have the potential to disrupt and delay projects. These activities may start as early as planning and approval stages on projects and extend right the way through them. Protests at Newbury and Manchester Airport have shown that the make-up of the anti-lobby is difficult to gauge. There are the well known professional protesters who use direct action. The groups who need more attention are those who have the resources to mount campaigns and have a political voice. In many cases these are communities fighting against projects. They mobilise the resources present within them to gain media exposure. Under such scenarios it is important not to adopt a confrontational approach but to stress the benefits of the project and to explain how concerns are being dealt with. The lesson at project level is that people naturally resist change, and the best way forward is information on how the project improves their quality of life. Any project that is perceived as not having benefits would require a 'spin' put on it to make sure a positive message goes out with it.

Corporate-level protests are difficult to assess. In many cases they are the result of dissatisfaction with responses given at project level. Such protests are made because there has not been a satisfactory response from senior management. In extreme cases there may be picketing or boycotts of products and services. All these present negative images of the

and act to counter any misinformation about it. The organisation has to be seen and projected as different from the image that is being portrayed. At times when the organisation is under pressure those groups with whom the organisation interacts can be a useful counterbalance. The track record of the organisation makes a difference as to whether people believe the counter-arguments. Actions speak louder than words.

The way forward

Does the adoption of a community policy mean massive new investment and cost to the organisation? The answer is no, because most construction organisations are involved in some or all of the following activities:

- Donations to charity
- Deployment of staff on community projects
- Environmental schemes
- Helping the unemployed
- Helping the disabled
- Educational links with schools, colleges and universities
- Open days and visits to corporate offices
- Visits to project sites
- Information aimed at local communities.

Construction organisations are therefore already involved in activity that could form part of a more structured community policy. The expenditure is already taking place on community activities but a cohesive and strategic focus is missing. The way forward would be for companies to rethink their approach to these activities and provide a sharper focus which will help them to improve their corporate image and that of the industry as a whole.

Summary points

- Communities in which we operate are critical for success.
- Community relations reinforce the norms and values of the organisation.
- Construction community relations can be approached at two levels, corporate and project level.

- Construction community relations should be based on a focused approach that ties in the business goals of the organisation. It should have its basis in a business case.
- Companies need to act responsibly and this needs to be communicated.
- Every action that is taken should be designed to have a positive impact on the local community and to improve the quality of life for everyone.
- More clients expect organisations to adopt a community approach to projects but companies need to ensure that the client has attended to all community issues before the company starts work on a contract.
- Interact closely with the client from day one to be aware of all community issues. The client will have been aware of all issues raised during the planning process whereas the contractor is new on the scene and has no idea what opposition there is or from where it may come.
- Have open days at the local library before you start work and then midway through the project, to allow locals to talk to the engineers and ask what will be happening, etc.
- Construction companies should be proactive and appoint project liaison engineers and organise forums with local people and other interested parties.
- Construction organisations should not underestimate the power of community action.

Keeping Our Own House in Order

Introduction

At an internal meeting of public relations managers representing more than 12 group operating companies of a large European construction group, the group public relations manager told his colleagues that, as public relations practitioners, all present should communicate as often and as much as possible with their staff. Whilst all nodded sagely in agreement with this remark, one manager set the room alight with argument when he countered: 'I agree with the sentiment, but you should not communicate simply for the sake of communicating; there needs to be a reason for communicating and a need for proper messages to be communicated. Without these, communication will be counterproductive.'

Both views are of course correct. But both views need to be merged to ensure that there is enough material to communicate and that the communications are actually achieving an objective. If there are masses of communications containing uninteresting, superfluous or patronising material, there will quickly be apathy or even complete disregard of the medium employed.

Good communications are of fundamental importance in the construction industry where clear and consistent messages need to be managed in support of business objectives. With a large, mobile workforce that is expected to travel and to move on from one site to another, effective communication is one of the strategic tools available for gaining employee commitment; improving morale; increasing productivity, quality and safety; and introducing new technologies.

The pace of change within the industry today makes internal communications even more vital for business success. Only companies that respond fast to changing commercial opportunities and working practices will survive, and only those who can communicate those changes internally and motivate changes in their employee attitudes and culture can respond fast to such changes. Thus communication and those responsibility for internal communications are integral to the strategy development of any construction company.

Moreover, the staff newspaper, e-mail, noticeboards and monthly management meetings are not the only issues that are central to successful internal communications. In construction, more than in most other industries, it is people and their own ability to communicate that are essential. Unlike most industries in a rapidly changing business world, the construction business remains essentially hierarchical. As such, line managers need to be equipped to transmit organisational and operational messages effectively and relevantly to other on-site and off-site staff and subcontractors. Thus senior executives need to be involved in internal communications to reinforce the vital role that communications play in not only creating but also driving the vision and strategy of a company.

Internal communications as part of the development of the corporate culture

In Chapter 11 we will see how vital internal communications can be in enhancing a corporate culture centred on safety. But the same applies to a construction company intent on enhancing a culture of quality, environmental awareness, or any other forte that gives a company a fully professional outlook on its work practices.

Moreover, technology, innovation and experience are all critical to an organisation that operates nationally or internationally but that is managed by geographical or sectoral areas. For instance, the knowledge gained by a project manager, his agents, sub-agents and section engineers on a contaminated land rehabilitation project in Motherwell can, if the internal communication system works, be used to great effect in winning and then successfully managing a gasworks reclamation site in South London. Or a novel technique of piling developed by an operating company in the Netherlands can be adapted by a sister company in Germany to win a major maritime engineering project using the technique as an alternative bid – if the know-how is successfully transmitted internally!

Hence we have several levels or areas of internal communications to address within the construction industry. The first is straightforward employee communications to transmit management messages and corporate culture. A second is operational, organisational communication to keep a company operating cohesively and with shared objectives and cultural aims. Another is internal marketing communications, where technology information transfer can be used for commercial or confor our internal communications: the diverse, multi-disciplinary nature of the workforce and the transient geographical nature of their existence. How do you communicate simultaneously to the area director, the young graduate civil engineer, the dumper driver and the accounts department supervisor? How can you communicate with the 9.00 a.m. to 5.00 p.m. office secretary in the same way as you do with the Portakabin-based engineer whose daily work is governed not by the hour but by the tide, the wind factor and the length of the piles to be driven in that sector? And how can you make what you say interesting and informative to all of them?

The answer is that such a task is impossible except through a variety of different media, each customised to a target audience. These include induction packs, company manuals, staff newspapers, noticeboards, e-mail, social clubs, the internet, external publications, internal technical visits to site and technical seminars, documents relating to specific issues (safety, environment, quality) and so on. It is through the managed direction of this pot pourri of media that successful internal communications will be implemented within a construction organisation.

What do we want to achieve?

So, what are we trying to achieve in undertaking all this communicating? The main objective is to motivate and inform our staff in order to help achieve our business goals. Construction is not so much about tower cranes, traffic cones, bulldozers, bricks and tarmacadam but about people. Our business depends upon people and their willingness and ability to meet the objectives established by the company.

In this business it is easy to feel isolated as an employee, particularly in the civil engineering sector. For example, you are a young engineer based in a collection of three Portakabins just above the high water mark on the coast of East Anglia with a foreman and 16 operatives for company day in, day out, with working hours depending upon the tide and the delivery of rock armour by barge from Norway. You may have a group of hostile fishermen breathing down your neck because they cannot fish for lobster and crab within your two-mile exclusion zone, and kids keep on vandalising your cabins, equipment and fencing. You get home some weekends and your area contracts manager visits you once a week. So what is your problem?

Your problem is that you want to belong, to know that you are contributing to the firm and the organisation and that they appreciate your contribution, they support you and they are interested in you. You want to know that you are not the only one in the company out on a limb. You also want to know what all the other young engineers are doing, how the company is performing, what else is going on, what the latest situation is with regard to training, what other contracts are being won and where. You also want a bit of guidance, some reassurance, some strategic messages, and maybe even a pat on the back.

The main aim of internal communications therefore is to keep that engineer happy in the job and to answer all those questions (or as many as you can) that he or she is asking. Only by answering those questions will the internal communication system be successful because otherwise you will be unable to promulgate those equally important *management* messages that are essential to a company's strategic plans.

This leads us on to strategic management messages. Another key to effective internal communications is meeting the 'corporate objective': to give out serious management messages for future corporate development. This could be through any number of media, and concern any aspect of corporate strategy. It could be a safety poster and accompanying safety newsletter warning employees of a change in the law relating to contamination of groundwater, thus alerting them to the need for vigilance in terms of diesel spillage or chemical wastes. Or it could be announcing new areas of training made available through the company's moves towards 'Investors in People' accreditation.

The 'Investors in People' or IIP scheme itself is indeed part of the motivating force behind many construction companies in broadening and improving their internal communications. And if any scheme pointed out the objectives for communicating within a company, then IIP certainly leads the way. IIP was launched by the Government in November 1990. At its heart is the National Quality Standard for effective investment in the training and development of people to achieve business goals.

Gaining recognition as an 'Investor in People' under the Standard is seen as a way of reinforcing a company's commitment to continuous development of a qualified and motivated workforce. Moreover, the Standard provides a planned approach to setting and communicating business goals and developing people to meet those goals. Thus, although based essentially upon staff training and development, IIP represents a company's commitment to communicating and caring for each member of staff and every employee.

Such guidelines or expectations endorsed by IIP are made manifest in the need to inform all employees about the performance of a company or a group. Obviously, 'Commercial-in-Confidence' rules apply, but a company's figures in terms of current and projected turnover, profitability

They want to know about turnover in each sector, region and profit/loss situation; they also want to know about the performance, in a larger organisation, of sister companies or other divisions.

What tools and techniques can we use?

All of these objectives need to be met by a wide variety of communication tools. No single medium can meet all of these objectives. However, the most effective, well tried and meaningful means of communication in any industry, but particularly in the widespread construction industry, is personal contact and word of mouth. And this needs to be managed otherwise it takes the form of rumour or tittle-tattle. Thereafter, from the initial induction programme for either a graduate or an operative, internal communications then takes on the more common forms of staff newspapers, videos, noticeboards, manuals and the like. The following brief sections simply identify the types of media through which we can, as an industry, communicate with all of our staff effectively and constructively.

Induction packages

It is vitally important that any new recruit, however senior or junior and whether a professional, a specialist engineer or a site labourer, needs to be *welcomed* and made to feel part of the organisation from day one. The welcome should not be a patronising 'Here's the company brochure, welcome to the site, here's the shovel, there's the hole!' It needs to reach each particular audience that it addresses. The more customised the induction is for the recipient, the more effective it will be, and the greater the impact, the sooner will loyalty be generated.

The construction industry is in fact one of the most advanced industries in the context of its induction programme. This has been driven largely by the tremendous pressures exerted upon the industry by the health and safety institutions over recent years. The need to communicate safety from the minute a person enters a site is paramount. Therefore, the conduits for communication have been established very strongly at the very roots of the industry: the sites themselves (see Chapter 11: Communicating safety in construction).

However, this grass roots safety induction regime that has become imperative on every site for every construction company has not always been matched by many induction programmes for more senior staff. Company familiarisation is still very much a take it or leave it situation in some companies. With the threat of shortages of qualified staff, only those firms that offer a serious, well thought out induction programme can count on a degree of staff loyalty and therefore on staff retention. This is particularly the case with graduate recruitment and induction. It is quite common for young, freshly graduated engineers to move to two or three different construction companies in their early years, to experience not only different types of work (civil engineering, tunnelling, building, geotechnical and other specialisations) but also different types of working conditions and to find, different opportunities for advancement. If these young graduates are given a warm welcome and there is an early, genuine effort to communicate the ethos and culture, direction and management objectives and other messages about their new employer, the company will be rewarded in the long term.

One of the leading construction companies in the UK has recently introduced a 'Personal Induction Guide' manual which describes itself as 'a comprehensive tour of the company, the people, the structure and how it works'. It is a state of the art tool developed by the company's own human resources department in conjunction with a college of technology and drawing upon both industry and academic expertise and experience. It includes not only a very simple 21-page booklet (there are 42 pages, each reverse page being for note making with reference to the facing page), but also a personal induction PC floppy disk. The induction manual and disk operate in conjunction and encourage the reader to interact with his or her PC and the manual as well as other literature included in an overall starter pack. It is simply a guide that takes any new recruit upon a quiet yet detailed journey into the company that he or she has joined. But it is also fairly demanding and practical. It asks the newcomer to check various elements, to read certain literature, to fill in checklists, and to discuss issues with the line manager. Finally, it gives a certain amount of early instructions and ground rules essential for 'survivability' in the company.

The Induction Guide in question begins in the warmest and most reassuring manner possible, explaining exactly what this small package is and what one is meant to do with it:

'Welcome to the company! You may well find yourself already allocated to a specific site, area office or head office and this Guide has been designed to help you to manage your own induction programme with the support and guidance of your manager.

There are action boxes highlighted throughout the Guide and a disk to provide additional illustrations. Follow the instructions to use the disk in any Personal Computer. The Guide will refer to literature in the

access to the information. There are also instructions about the forms in the Starter Pack and a checklist at the back of this Guide.'

The Guide sets out from the start to be user friendly and yet totally practical. For the average office worker, accounts department clerk or IT assistant, the Induction Guide may seem irrelevant, but if one joins the Company as a civil engineering graduate, technician or construction-specific member of staff, then the Guide exudes practicality in its intro-duction:

'Line Manager's local induction
On your first day, you should be able to meet your line manager or someone delegated to help you. Clearly you will need some guidance about the local arrangements to help you fit in to the work team and its operations. The items below are a guide to help you and your manager cover the essentials to ensure you operate safely and confidently in your work location.

Local health and safety arrangements (to be completed on arrival):
- Emergency procedures (including fire drills and safety induction)
- First aid arrangements

Local domestic arrangements:
- Toilets
- Security systems
- Catering arrangements (if any)
- Car parking arrangements
- Local working hours
- IT systems and availability
- Telephone systems/fax
- Stationery

Introduction to local workplace:
- Guided tour of location
- Introduction to operations and activities
- Local organisational structure
- Workgroup members – roles and liaison
- Administrative services and support
- Social activities (if any)'

The Guide then proceeds with a history of the company, a review of its 'culture and objectives', its quality standards, structure and policies, procedures and conditions. Interestingly in the context of this book, under the chapter heading 'Head Office Structure' the Guide gives all new employees the following briefing about the role of the Public Relations function:

> 'PR is responsible for the company's reputation, corporate identity and culture. The department supports communication between the company and clients, shareholders, employees, the media, the construction profession and the general public. Its main responsibilities also include... **Important note:** should you have occasion to have any contact with the media, company policy is that all such contacts must be referred in the first instance to the Head of Public Relations or, in his absence, the Managing Director.'

Company manual

Most construction companies have the official company manual which contains the most critical communications a company has with its senior staff. Generally speaking, this manual has the full authority of the Board of Directors and is mandatory for use throughout the company. It relates to all procedures within the organisation, from health and safety, tendering, design, purchase, plant and insurance to quality, environment, cost control and public relations.

How the company operates depends largely upon the procedures laid down in the company manual. In this respect the document is probably the most important communications tool within an organisation. Proper production, distribution, control and management of the document are equally critical.

The most effective management of the document is through targeted distribution to those responsible for managing operations: site agents, safety officers, department heads, contract or project managers, commercial managers, and the like. It needs regularly updating to reinforce awareness among all recipients of the importance of the document. The contents themselves are invariably 'commercial-in-confidence' and are particular to each company. Thus it is not within the realms of this book to advise on the contents of the company manual, but simply to stress the overwhelming importance of the medium for essential internal communications.

Staff newspaper

As with most industries, the in-house staff newspaper remains the main medium of internal communications for most construction companies. It

... flexible media in its ability to get numerous messages over to and understood by all levels of employee. Finally, it is, or can be, relatively cheap to produce.

However, one of the most common reasons for staff newspapers becoming a complete failure in the eyes of the recipient (the *potential* reader) is that the publication has been written, designed and produced simply for the sake of it, and not in a defined manner with specific objectives and planned messages. As we pointed out in the introduction to this chapter, staff newspapers should be used to communicate the corporate culture to all employees and not as word processor fodder backed up by regurgitated, boring, site-progress photographs simply to pay lip service to the accepted need to be seen to be communicating with staff.

On this basis it is clear that such an important task should be managed and controlled by someone or a group of people or an outside agency with specialist knowledge and expertise in internal corporate communications and particularly staff newspaper production. The managing director's secretary, however eloquent and however proficient at the PC keyboard, may well be unable to produce a staff newspaper or even a newsletter that is worth the paper it is written upon. People have to be trained and need a solid grounding in this sort of work to make the necessary editorial decisions so vital in the production of an effective staff newspaper.

Some problems specific to the construction industry in terms of staff newspaper production arise out of the very character of the run-of-the-mill site engineer. Often engineers, for all their brilliance with the theodolite or their understanding of the characteristics of the setting conditions for specific concrete mixes, or of the strength of a rebar cage, are unable to write clearly. They can quite eloquently and rationally explain orally what they and their team are doing in the construction of a bridge abutment, for example, but if asked to write it down will produce either two lines of unintelligible gobbledegook or a five-page technical dissertation that is understandable only by a qualified graduate in civil engineering.

By the same token, the potential readership is made up of such people. If they often cannot write intelligibly, can they be expected to read the flowery, often patronising and simplistic verbiage that many a staff newspaper contains? The answer is probably not. Thus the newspaper must be written and presented in a way that will appeal to its readers. This means not just the style of writing but also the presentation and content of the text material. There has to be a balance of coverage to make interesting reading for the broad range of readership within a construction company. In particular there needs to be sufficient information on the company itself, a selection of current contracts of particular interest, a balance of coverage for different geographical or divisional sectors, some fundamental corporate core messages (performance figures environmental or safety matters, training procedures and such like) and then an element of the social and sporting activities of employees in and out of the work environment.

Tarmac World, the staff newspaper for Tarmac employees everywhere in its worldwide operations, shows how well the construction industry can compete with any industry in the production of a good staff newspaper. During 1997 *Tarmac World* won two major publishing awards. Not only was it voted Newspaper of the Year in the Corporate Publishing Awards, beating off strong competition from the likes of Railtrack, Lever Brothers, Marks & Spencer and National Power, but it also won the supreme Gold Award in the Communicators in Business Awards (CiB), the biggest of their type in Europe.

Tarmac World was redesigned in mid-1996 in line with the Group's new corporate identity (see Chapter 2: Corporate Identity), using an outside agency that specialised in internal communications and staff newspaper production. It provides a good example of the necessary mix of content for a construction industry newspaper. Produced in full colour and generally around 20–24 pages in total, *Tarmac World* is a monthly newspaper emanating from the Group Communications Department of Tarmac plc. The contents include 'News' (a section that includes contract awards, openings, topping-out ceremonies, special achievements, and new services or materials within the Group); 'In My View' (a think piece from the Group Chief Executive that looks at such issues as the setting up of the Major Contractors' Group and the establishment of a new Group Charity campaign); 'Letterbox' (a selection of staff letters); 'Features' (often project or contract specific); 'Family Focus' (including sporting, social and general interest features and stories); 'People in the News' (retirements, long service awards, obituaries); 'Sport'; and 'The Environment'.

In terms of core company or group communications, several key messages appear in every issue, albeit within the user-friendly environment of other, more relaxed articles. For example, the April 1997 issue of *Tarmac World* announced higher operating profits for 1996 with messages from the chairman, Sir John Banham, and chief executive, Sir Neville Simm; a full description of the group's new operating structure and divisions; a feature about Total Customer Service and the importance of customer relationships; and then some key environmental pieces. Other issues contain specific messages about safety, the reorganisation of a division or the group's participation in various industry forums.

By way of a contrast, the same newspaper gives a wider appeal to staff

Tarmac World is the group-wide staff newspaper for operations and divisions worldwide. Other groups have different corporate approaches to staff newspaper production. Some will have corporate staff newspapers for each country in which they operate, with, for example, core group messages appearing under one banner in Germany, written in German and with specific news and information about German operating companies in particular, but with other articles covering operations elsewhere in the world. The same core group messages will also appear in Dutch for the operating companies in the Netherlands, and in English for the UK operations and staff. Each edition is customised to the particular national audiences, cultures and businesses but retains the same group messages whether the reader be Belgian, French, Dutch or German.

In the same way, many companies that operate within a larger group, whether nationally or internationally, prefer to keep their own 'divisional' or 'company' newsletters or newspapers in addition to the overall group newspaper. This need not be a problem, so long as it still retains the overall group culture and objectives. On the positive side, it reinforces the sense of belonging of employees to a particular division or company within a large group, and can help with specific business objectives.

In terms of staff newspaper circulation, there are conflicting views about the necessary distribution methods. Often, the distribution of the staff newspaper is organised through the 'dump it and they'll pick it up' method. A pile of in-house newspapers is left in the office reception and a small pile appears in the site canteen every so often. If the papers are picked up and read, that is fine; if not, the pile simply goes into the skip when the next pile arrives. Obviously, this system is not really an exercise in staff communications. But it is amazing how many staff newspapers are 'distributed' in such a fashion.

More common is an internal staff distribution system where each member of staff is specifically sent a copy to his or her place of work, individually named on the envelope or cover. However, in the construction industry, where employees can be moved on from site to site several times a year, even from country to country and from division to division, the best realistic solution to staff newspaper circulation is to mail a copy to each employee at their *home address*. Not only should this guarantee that each member of staff receives a copy of the newspaper, but it also includes the employee's family – reinforcing the sense of 'belonging' to a company or organisation – a sense of 'caring', even. If the newspaper goes home, it should show the family that the organisation needs to communicate not only with the individual employee but also with his or her whole family. There are issues which affect the welfare of all, so the messages should appeal to the husband or wife of the employee or to other near relations.

Another issue concerns frequency of publication. Generally, staff newspapers appear monthly. This gives enough opportunity to generate real stories, real messages for staff and effective coverage of geographical or business sectors in an average-sized construction company. Any more frequent than monthly will give rise to the need to produce material for the sake of it. On the other hand, a staff newspaper produced only quarterly may not be able to communicate effectively with employees. However, as long as the contents are well planned and are meaningful to employees, even quarterly newspapers can be the most effective means of corporate internal communication.

Other publications

Staff communications envelops a whole range of publications apart from the staff newspaper and the variety of new staff induction literature. It includes pension documents, safety publications, quality control documentation and other news sheets. It also includes most *external* marketing communications: the company's brochures aimed primarily at existing and potential clients.

It is well known that one of the strongest reasons for advertising in the motor vehicle sector is to reassure existing marque owners that their choice of vehicle remains at the forefront, is respected, is top quality, and so on, to make the existing owner feel good and feel proud of his or her car. Secondly, it may make others want to purchase a similar car. So with employees. If they see their company up in lights, with powerful publicity in advertisements, brochures and magazines, then it reassures them and gives them pride of 'ownership' of the company.

Thus the company's external magazine, aimed specifically at present and potential clients, architects, consulting engineers, investors and other outside influences, is also a powerful internal publication. Try leaving your latest marketing brochures in the reception of your head office or regional office and just see how many employees pick it up and read it avidly or take a copy away to read later and keep. It is talking about *them* and their company and is therefore about who they are and what they do. It allows others to understand them more.

Conversations down at the local usually at some stage include the question, in one form or another, 'And what do you do?' Knowing what the company stands for, its services, its experience, track record, expertise, geographical spread, quality, innovation and so on, all contribute to

cerning specific sectors or divisions or the whole organisation, are therefore a vital tool in internal communications. Do not forget that every employee is a salesperson for the company. The more that they can say about the company, the greater the information flow and the higher the profile. Thus it is good policy to make all marketing and external publicity publications available to staff, though not necessarily distributed directly to every employee.

Other vital publications for staff concern safety. Chapter 11 shows clearly how communicating safety in our business is of fundamental importance to the whole business of a construction-related company or organisation, as well as to every individual employee. There is no harm in repeating how important it is to issue regular safety bulletins promulgating the latest safety issues and regulations. These could include posters, videos, talks and other communications packages. Most importantly, every member of staff should receive a full copy of the company's health and safety policy booklet.

A company publishes a wide range of other documents for staff. All of these, whether they be booklets explaining quality management systems and procedures or the company's environmental procedures and regulations, are vital cogs in the wheels of employee communications. Most important, however, are personnel department publications. These concern pensions, additional voluntary contributions (AVCs), training, staff development, medical schemes and other issues concerning a person's employment and terms and conditions. All of these need, first and foremost, to be staff communication tools. Input from the public relations function to this range of publications is essential to ensure that they are technically accurate and effectively communicate the messages required.

Verbal: presentations

Nothing can beat the face-to-face staff communications method, from the top to the bottom. We have already discussed the staff induction programmes, ranging from the graduate induction three-day presentation from department heads at head office to the labourers' 'toolbox talks' given by the section foreman on site. However, there is a myriad of other face-to-face communications scenarios vital to the successful promulgation of internal messages.

Some companies, or at least some major divisions of larger groups, have an annual 'roadshow' whereby senior staff get an opportunity to address

All or most staff at least once a year, not only to give information but also to receive feedback, i.e. to communicate.

For example, one contractor which, like most large construction companies, is organised on a regional or area basis has an annual 'address the nation' type of event that is organised to coincide with the publication of the group's annual results. At this time, every area or regional office provides a venue (usually a hotel or restaurant environment) to invite all area staff for a formal presentation of the company's performance in the previous year and of objectives for the coming year and for long-term development.

The managing director then spends an evening in each area office. He explains the group performance, turnover and results, the individual operating company or divisional turnover and results, and then the specific results of the particular area, region or division being addressed. He looks at the positive elements, pats the appropriate backs, and reveals shortfalls. Most importantly, he then outlines the company's strategic objectives and plans for the coming year.

Then the presentation moves on to a sort of instant snapshot of the company at that moment in time. There is a review of key existing projects being undertaken, and lessons learned from recent contracts are discussed. Finally, there is a review of other operating companies not directly connected but also part of the overall group structure.

After this formal presentation, lasting approximately an hour and a half, staff are invited to question the managing director. Whether the questions be area or division specific or of a more general, strategic nature, they can be addressed directly to the top man.

Although this event only happens once a year, it allows every member of staff to be fully involved in the company's overall strategy. It allows everyone, be he or she section engineer, site secretary, ganger, foreman or divisional director, to assess the company's objectives and direction and then question the managing director personally. It is very much a two-way communicating event.

Not every company can do this sort of roadshow. Often the organisation is simply too big. But perhaps it could be undertaken on a divisional basis if this is the case. It would certainly pay dividends in terms of staff commitment and trust in management where such an open-book management style is demonstrated.

Moreover, such an annual event can then be duplicated if necessary further down the line and perhaps on a more frequent basis. Many regional or area offices organise their own monthly or quarterly management meetings. These can be arranged on similar lines. Individual managers can then adapt similar events at site or contract area level. Thus a 'cascade' of face-to-face talks from top to bottom and vice versa can be show as described above with the chief executive, managing director or chairman talking to camera or to an interviewer about results, objectives, structure, strategies and such like. This is then circulated to area offices, divisional directors, heads of department or other management levels and it is left to them to undertake their own presentations and feedback. But this cannot really substitute for the real thing.

Another type of verbal presentation undertaken by some companies takes the form of technical talks. One company in particular organises fortnightly technical presentations at head office whereby project managers or agents are invited to come to head office and undertake a half- or three-quarters of an hour presentation to head office staff (or central services staff including design, planning, engineering, procurement, personnel, finance, insurance, marketing and other service sectors) about a particular project. This is a vital form of communication as it transfers 'coalface' on-site real-life situations directly back to the office-based support services staff. It gives them a chance to understand their role in direct input to a site's operations. Again, it provides that vital two-way flow of communication between operational staff and support service staff. Only such communication can alleviate misunderstanding or lack of co-ordination.

Finally, many construction companies organise their own internal technical visits. Where a particularly innovative or interesting project is being undertaken and where it is felt that other employees not directly involved may benefit from understanding those particular works, then technical visits are organised and a party of staff is invited to attend. This usually involves a formal presentation by the project manager and other engineering staff on site, followed by a full tour of the works. Participants could include central service staff or other site-based operational engineers who may benefit from an understanding of the projects in hand.

Such technical visits are another vital link in the overall staff communications package or network. They provide mutual understanding and respect among staff and peer groups within a company. They also promote a greater sense of family and the sharing of knowledge and information among members of that family for mutual benefit.

Other communications

A wide range of other employee communication methods are available to a construction company and are equally important in the overall mix of media that can be deployed. E-mail and the internet are two of the most

a hugely growing multitude of communication. With every site having PCs
or at least a laptop computer, the provision of modems and inter-
company e-mail networks is a growing phenomenon. These are particularly useful for the transmission of large documents or volumes of
information. However, e-mail is dealt with elsewhere in this book.

One of the quickest and most obvious ways of assessing the success or
otherwise of a construction company's internal communications system is
to visit a construction site and look at the noticeboard. Is this covered in
bed-and-breakfast/lodgings advertisements, press clippings from the
local papers about the disruption caused by the construction works, the
odd health and safety poster that is the statutory minimum and a few
smutty cartoons relating to the social life of a quantity surveyor? Or does
it have a series of the latest company press releases, details of the organisational structure of the site itself, directions to the local accident and
emergency unit or hospital and a selection of the latest safety posters,
safety newsletters and organisational instructions?

A well kept and constantly updated noticeboard can be one of the most
important means of reaching the widest range of staff, both at area office
level and at individual sites. A mechanism for the distribution of
important documents to every site and a specified system for displaying
that material is essential for the proper use of noticeboard material. The
best way of ensuring that no one looks at the noticeboard is by covering it
with an array of dog-eared, out-of-date circulars. A board that is regularly
updated and contains important and frequent messages from the site
agent or project manager will ensure attention.

It should also be a matter of policy that any press release issued from
the public relations department should be sent to every site and displayed
on noticeboards. Not everyone has access to *New Civil Engineer, Builder,
Construction News* and *Contract Journal*. So this mechanism helps inform
staff of the latest contract awards, staff appointments, new technologies
and such like.

The social club is another means of communication *between* staff as
opposed to *with* staff. Any interaction, particularly social, between staff
can help enhance the sense of belonging within a company. Even if it just
means 20 members of staff meeting once a year at the weekend for a few
rounds of golf, and dinner, bed and breakfast, the camaraderie that can be
found at such an event can produce dividends for the rest of the year.
Thus, it is in the interests of any company to encourage and support the
activities of a sports or social club. This could be either office based or
geared up to an area or region of sites.

Other internal communications are covered primarily by the person-
nel/human resources department with circulars on staff development,
training, pensions, contribution elements and other aspects of staffing,

the organisation. Here the success or failure of the system often depends upon the effectiveness of the internal mail mechanisms. If a fax takes six hours to get from the post room to the recipient two floors away, or if an internal memorandum takes three days to get from Cardiff to Glasgow, then the mechanism has failed. Likewise, if a staff notice takes four days to get from head office to site noticeboard, then the system has not worked.

Summary points

- The successful internal communications of a large construction-orientated organisation are critical to the effectiveness and success of the management of that company, and therefore to the success of the business itself.
- Communication of key corporate messages and the promulgation of a company's culture, whether about safety, quality or even environmental aspects, are fundamental to how staff work and perform and to how much they commit themselves to the common cause. A company's very cohesion is totally dependent upon its internal communication.
- The main objective is to motivate and inform company staff in order to help fulfil the overall business goals.
- A key to internal communications is to meet the corporate objective: to give out serious management messages for future corporate development. This could be through a variety of media and could concern any aspect of the corporate strategy.

Corporate Communications and the Role of the Project Manager

Introduction

Construction activity revolves around projects. There is an inextricable link between the construction project and the survival of its parent company. Any organisation that sees construction as its core business relies on projects to generate profits, to act as a 'cash cow' and to provide an outlet for its technical capability. The outcomes of projects are central to business success in the construction industry.

The art of project management revolves around the artist, the project manager. He or she is the focus for the success of a project and is accountable if it fails. There is considerable responsibility for the designated project manager as he or she is charged with success or failure. This role may be a general responsibility for a project or at the level of site management. The chapter focuses on the need for managers to develop the public relations skills that can make a significant contribution towards improving communications at the level of the project.

The project environment and its stakeholders

In an earlier chapter we described the importance and influence of stakeholders for the business of the company. This approach can be applied equally to projects. Project stakeholders are those groups or individuals who believe they are affected or can influence projects in either a positive or a negative manner (see Fig. 10.1).

The pressures on project managers have increased as sponsors, government agencies and the public become more aware of the impact of projects. Failure to recognise the existence and potential power of project stakeholders can have serious implications for the project as it moves into more detailed implementation.

The project environment is normally categorised as having 'internal' and 'external' environments. Internally are those people who have direct

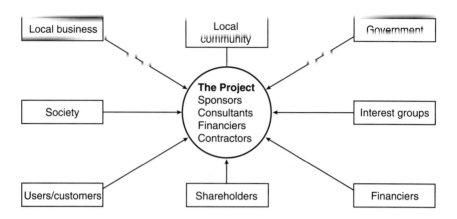

Fig. 10.1 Construction project stakeholders.

responsibility for projects and are in theory controllable and predictable. These stakeholders would consist of project sponsors, owners, project team members, project champions, consulting engineers or architects (Turner, 1995). There does, however, remain sufficient scope for internal conflict as many of the parties may have conflicting objectives for the project. Project communication internally is a vital element for the successful facilitation of the project.

The 'external' stakeholders are free to behave in any way they choose and need not have any regard for or loyalty to the project. These groups have the potential to galvanise both political and public support in either a positive or negative manner. The anti-roads environmental campaigners are a good illustration of this point. They have been able to galvanise opinion and to claim support from social groupings to an extent that would hitherto have been unthinkable. Stakeholder assessment has thus taken on a greater degree of importance, and a planned approach is needed.

Project strategy

An important part of any project is the strategy that is adopted. Project managers will develop strategies to cover contingencies that may occur over the life of the project. It would be sensible to adopt a similar considered approach to stakeholders as they represent a major force for possible disruption of the project. Using the ideas of stakeholder identification set out in Chapter 1, the first step is to identify the project

group. Stakeholder behaviour w̲... .

which are created by the project environment or by external issues. Research has indicated that project success is dependent on good communication (Baker and Murphy, 1988; Wilerman and Baker, 1988).

In order to extract the correct level of information a PEST analysis should also be conducted. PEST represents the political, economic, socio-cultural and technological influence that project stakeholders may exert. The analyses enables the project manager to identify the stakeholder and their stakes and also the main thrust of their influence. In order to plan a successful project relations campaign it is important never to under-estimate the level of influence. The project manager has to adapt the project strategy to take into account the influences of the stakeholders. This can only be done by integrating elements of stakeholder influence into the campaign.

Developing new project management skills

The skills of the project manager have been the source of much argument. However, it is recognised that certain attributes and skills are needed to perform effectively. It should be noted that these attributes are not the exclusive domain of project management. The common skills in the Project Management Body of Knowledge are (American Project Management Institute, 1996; Association of Project Managers, 1996):

- Leadership
- Technological understanding
- Evaluation and decision-making ability
- People management
- Systems design and maintenance skills
- Planning and control skills
- Financial and commercial awareness
- Procurement skills
- Legal awareness
- Negotiating skills
- Character: enthusiasm, drive, determination, vision
- Communication.

These skills can best be summarised as an axis between technical, legal, administrative and people skills. They provide a sound basis for the achievement of project objectives. The development of the various bodies of knowledge are about achieving project objectives within cost, on time, and to the required level of performance. These project objectives form the basis of success or failure for project managers. The well-known American project management educator, Harold Kerzner, adds good client relations as another objective of project management (Kerzner, 1992). Good client relations is a fairly broad statement and could hold a variety of meanings. For our purposes it should mean delivery of the project to the client with a minimum of fuss or bother to that client. It is with this view in mind that new dimensions are needed to project management. Good client relations is also reflected in good community relations during the course of construction since the client has to maintain a presence in the environment well after construction is completed. Good local relationships can be destroyed by a bad contractor or project manager.

Murphy's law

Project managers make use of the variety of skills they require to plan, implement and control projects. They are able to provide leadership, draw teams together, negotiate difficult issues, and resolve complex problems. Construction projects do not always go to plan, and the project manager is often in the firing line. It is at this point that the skills of negotiation, leadership, decision making and communication come to the fore. These are the 'soft skills' that are crucial when things go wrong but are often neglected in the search to meet time, cost and performance goals. Murphy's law, 'If things can go wrong they will', applies to projects. Communication and public relations techniques come to the fore when things go wrong. And they can also be maximised when things go right!

The project cycle and project management interaction

Every project goes through a cycle consisting of a series of different phases. Typically these phases comprise the following:

(1) *Project initiation:* A project is likely to be initiated when there is a prediction that there will be a demand for goods and services produced as a result of the project. At this stage the promoter of the project may sanction the use of resources to investigate and appraise alternatives to achieve particular objectives.

determine ~~whether the ten~~ returns, novelty, risk, etc. have been made. The results of this stage are specific: the project is sanctioned or rejected.

(3) *Design development:* Design ideas are generated during this stage, and costs estimated and evaluated. Decisions relating to quality must also be made during this phase and are directly tied into design decision-making. Technical issues including buildability may also arise out of the development of design. Decisions taken during this stage will influence the level of resources required on the project.

(4) *Project planning:* At this stage detail programmes are produced to identify critical paths and purchasing and other issues relating to the mobilising of resources to a set timetable.

(5) *Project implementation:* This stage involves a series of physical tasks to implement the project. A great deal of activity associated with co-ordination of tasks and monitoring and control is undertaken. This is the stage when objectives are being achieved. The environment of the project is more dynamic as objective accomplishment becomes a central driving force. It is at the end of this phase that the project is delivered.

These stages are present in every project and the project manager will be required to deal with the interfaces at each of these phases. In construction project management the same applies, but may differ due to the start of involvement either from initiation or during other stages. The philosophy is still constant despite the starting point. Each of these phases will require the application of particular project management skills. On larger projects it may be possible to identify and separate activities during these phases to make management more effective. Some of the phases are related to mechanistic forms of management, while others are more dynamic in their requirements. The mechanistic phases will accommodate meticulous planning, slow change and structured communication and briefings. The more dynamic phases of the project life cycle will allow less time for all types of activity and require faster responses to activity.

Building project level public relations strategies

In dealing with the project the project manager has to re-examine the communication skills that are required. It is an accepted approach to the management of projects that a strategy is devised to fulfil the aims of the project. It would therefore be appropriate to develop a project public relations strategy alongside that of the main project strategy.

A considered approach to project public relations is to develop a strategy for the project as early as possible. The public relations strategy should not be developed in isolation. It should be developed alongside the overall project strategy. The public relations plan should assist in achieving project objectives. The development of the project strategy will raise issues, contingencies, influences and problems; the PR strategy may assist in resolving or easing these problems. It then becomes possible to adopt a more proactive stance on project influences.

Any construction project, particularly during execution, will require the management of information, the event and contingencies.

- *Information management* is concerned with the dissemination of a range of types of information to staff and to a variety of external audiences including schools. This portion of communication is controllable and can be developed externally. The primary purpose of this information is dissemination of news related to the project.

- *Event management* is an opportunity to show off the project in the best possible light. Such an event is designed to improve the project's image and to strengthen organisational and project branding. Once again there is an opportunity to control events.

- *Contingency management* on projects is for incidents, actions or occurrences that were not expected as part of the smooth execution and operation of a project. Occurrences such as major accidents, spillages, deaths, the discovery of relics, strikes, bankruptcy and occupation of the project fall into this category. It is impossible to plan for every contingency but sufficient thought must be given to establishing key protocols that will need to be followed when an unforeseen event occurs. For example, a major chemical producer pursues the following procedure if there is a spill or fire at any of its plants. It does not issue any statements until its own investigation team has done a preliminary assessment of what has gone wrong. Until this point all comments are made by the emergency services. This allows the company to issue statements based on factual information and to ensure that they are on a sound legal footing. All types of project communication can be managed if planning is done early in the project. Crisis management will be examined in further detail in Chapter 12.

The project manager's parent organisation has to provide support for additional public relations requirements. New resources will be required for these activities. They may be part either of the project budget or of the resources from a public relations department. If there is public relations expertise in the organisation, then it should provide support and training in developing the skills and ideas mentioned above. This support function should also extend to the use of in-house expertise and company intelligence. It is essential that a flexible resource of material to support project teams should be prepared, kept up to date and co-ordinated by some form of centralised function. This function is responsible for selecting teams to attend interviews and presentations. The organisation needs to adopt a complete public relations strategy that extends throughout the organisation right down to project level. This ideal means that everyone and everything the company does contributes to its image. Everyone, from the labourers on site to the managing director, makes a contribution to the organisation's image. This view means that projects are integral to corporate image.

Managing the internal environment

The internal environment of a project is often overlooked in communication and public relations planning. Those organisations and individuals that operate within the project environment are seen as controllable and as supporting the project. This assumption is flawed in that many organisations and individuals will have their own priorities and objectives that may not be directly related to those of the project on which they are working. There remains sufficient scope for conflict. Communication and internal briefing can be used to ensure that uncertainty is avoided and that the project is not taken over by rumour. The use of proper consultation where appropriate can ensure that participants of the project buy into the decision-making process. Developing a strong team or 'brand' identity can also help smooth working relations and ensure internal project support. This form of branding is developed further through internal newsletters, logos, etc. The development of a strong identity within the project team acts as a further focus for the project.

The following sections detail communications issues during the course of the project life-cycle, commencing with the pre-qualification process. This builds on the earlier chapter which looked at marketing documentation to support team presentations and interviews with the client.

Pre-qualification communications and the project management team

Clients require that those contractors or consultants who are invited to submit proposals be professional in their approach. They expect them to provide a full team composed of project and site managers including the project manager responsible for the proposed project, and others with extensive contractual experience. During the interview and presentation the project manager sets the tone and determines the culture that is likely to arise during the project. It is also important that someone at director or partner level be accountable for the company and be able to draw on his or her expertise in resolving any contractual difficulties.

Preparing 'the pitch'

Like most activities, preparing the pitch is important. Effective interviews and presentations depend on the company or practice gathering as much information as possible on the following:

- The client organisation and decision-making team, and their expectations.
- The client's products and markets, and problems they may be facing themselves.
- Competitors' strengths and weaknesses and any relationships they may already have with the client.
- The strengths and weaknesses of the organisation or practice, past or current relationships with the client, the value of the service to be provided, and experience of previous similar projects.

Generally, a client specifies exactly what information is required and in what format, how it should be presented and by whom. However, sometimes the pre-qualification process is left open to the bidder.

This process allows a company to compare its services against client criteria and identifies opportunities. This will enable identification of a competitive difference which needs to be promoted through background publicity and brochures, but most importantly through direct contact during interview and presentations to the client. Analysis of the customer, competitors and the company itself needs to be converted into a differentiating and ultimately 'winning' message. This message needs to satisfy the basic expectations of the client team that the company has the appropriate track record, a high-calibre and reliable project team, adequate financial backing and quality management.

company from its competitors. A serious attempt to understand the client's requirements is essential. The client organisation may be made up of different parties with priorities. Any pitch should be aimed not only at the funders but also at the end users. This can be critical depending on the influence of the respective interests in the final decision making. Ultimately the ability to put across the understanding with the other requirements of the project will allow you to differentiate yourself from the competition.

Presentation teams are drawn from different functions of the organisation, i.e. project managers, contracts managers, quantity surveyors and planners. This requires cross-functional co-operation and communication. Although representatives may be highly experienced technical managers, they may require some training in negotiation and selling skills to enable them to communicate more clearly the problem-solving abilities of the team and its ability to respond quickly to changes in a client's requirements posed in the interview situation. On projects that are design and build, there may be an additional requirement to integrate external consultants under certain scenarios. The project manager as team leader has to take a decisive role in such activity. All specialists involved in the interview and presentation should be speaking from the same script and be clearly aware of the message the company is trying to put over. Good internal communication is required to achieve this as everyone from project managers and directors to site agents needs to understand what message is being put across. Pre-presentation practice and questioning should sharpen the presentation and ensure that the team is 'on message'. It should also help iron out any wrinkles in the presentation. Despite this seeming to be common sense, project teams can unravel during a presentation and subsequent questioning.

Managing the team is a critical part of project presentation. Equally important is managing the message. The entire strategy should be based on getting the message across. However, it is likely that any presentation for a client will have a time restriction on it. It is easy to fall into the trap of not getting across the answer to the key question as to how you are going to satisfy client requirements. The trap of technology, where the entire presentation is based on technological ability, is another area to avoid. Technology may be useful if that is what the client wants, but this may not be the case. The impression created during the presentation should have its impact long after that presentation has ended. Any message that needs to be put across should finish within the stipulated time. Failure to ensure that this is always the case is a direct result of poor preparation and planning.

Points to remember

- Audio-visual equipment such as videos may enhance presentations, especially if professionally produced, but only if asked for! However, their use should not in itself distract from the main message.
- Identification of what clients are expecting, their criteria for selection, the specifics of the project and what competitors are likely to be offering, should be the basis of promotional efforts.
- Clients and their advisors are looking beyond slick and glossy presentations and want to hear from people involved from director down to site level. The content of the presentation and the response to questions are equally important.

The project manager and project team can be better prepared for the pre-qualification process. Many of the objectives set at this stage have to be fulfilled during the project. Increasingly clients expect the executors of their projects to provide a total project management service which goes beyond traditional cost, time and performance ideas to cover project communications and the management of the issue and of the media. Developing understanding at this phase allows for a better planned project.

The project: execution

When the project gets to the execution stage a lot of the work has already been done with regard to project communications. The information management, event management and contingency management schemes should have been put in place as part of the overall communications strategy. Much of the thought and creativity is in the planning. The communication planning process should have identified the possible contingencies that may arise on a project. For example, ground breaking, accidents and emergencies, pollution, complaints, customer interfaces, project publicity, site visits, handovers and openings may be some of the events that can be pre-planned. Each project will have a different emphasis. Alterations to existing roads would probably concentrate on getting information out to road users rather than, say, emphasis on dust control on a city-centre project. The point about these issues is that many of them can be planned before the physical performance of work.

Before the physical work starts, many of the communication protocols need to be established. A who, when and how approach for each project needs to be established. The 'who' relates to the person responsible for any communication on behalf of the project. This may vary because in the

admitted. A protocol of this nature could be resolved in advance and should tie into corporate policy.

When are statements for communication released from a project? The type of information that should be released needs to be determined early as should the point during the project at which it should be released. There is also the issue of how to respond to requests for information, particularly from parties outside the project, for example from the media in response to rumours about it. The response should follow the guidelines laid out in media management policy. These usually take the form that all media contact be referred immediately to the press/public relations manager. There needs to be a consistent message and it also needs to be controlled. Subcontractors need to be told that they are not allowed to make pronouncements on anything unless agreed by the contractor and the client. Responses to crises should follow guidelines set out in Chapter 12 on crisis management.

Site event management

The project manager often finds that there are opportunities for visits to sites. These may vary from a local school to city analysts and investors. The first rule of event management is that unannounced visits should be forbidden. Forward planning is vital as the site should be operating safely and at its best. Equipment and machinery should be checked, and the site should be as clean as possible. The following need to be considered:

- Reception and parking areas
- Accessibility issues such as wheelchair ramps
- Site safety requirements: provision of hard hats, high-visibility jackets and safety boots
- Toilet provision, litter bins
- Refreshment provision and area: beware of dietary variations
- A presentation room
- Suitable briefing material and take-away documentation.

Managing the event or visit requires other precautions. The site safety officer should play a major role in the event. It is essential to start any site visit with a safety talk stressing all matters affecting safety. All stewards should know what to do if an emergency should arise while the visit is in progress. Route planning is important to avoid any actual or potential hazards. It is also important to remember that groups are better managed when they are small. The optimum size of group is about eight. Larger groups should be broken into smaller, more manageable ones, each with an engineer alongside to discuss elements of the contract.

Preparing the format for a visit is important. It is useful to start with a description of the project and its final purpose. Remember to pitch any talk to the level of the group present and avoid the use of jargon, abbreviations and colloquialisms. Be selective in the information that is presented. During the visit, explanations should follow the same procedure. Encourage and be prepared to take questions as this breaks up the monotone of the speaker. Patience is a virtue during site events.

The impression that the visit creates is important. Ensure that as many people on site as possible know that a visit is taking place and that a degree of hospitality must be shown to the visitors. Ignoring visitors only creates the impression of hostility. Make sure that visitors never leave the site empty-handed irrespective of profile or age group. A package of information or some form of 'freebie' bag should be provided. The key here is that the visit should stay in the minds of those who have visited the site. More than anything the visit should reinforce positive images of the company and the project.

Summary points

The manager of the construction site has an important role to fulfil in the overall business success of the company. Involvement starts from the bidding process and continues throughout the life of the project. As a communicator the project manager has a far-reaching influence on the way the project is portrayed. The key lessons that need to be borne in mind are as follows:

- Planning is a critical function for site communication.
- Establish a communication plan with a contingency view.
- Internal project communication and branding should not be ignored. This will assist in establishing a more cohesive and effective construction team.
- Pitch the message to the correct audience. This means understanding who you are communicating with, i.e. their backgrounds, needs and interests.
- Ensure that everyone is 'on message' from the project team during interviews, presentations, events, visits, etc.
- Ensure that all written communications contain and support the correct message.
- Establish clear protocols for communication as thoroughly as possible

- Preparation is critical.
- Positive messages should be generated about the project.

Remember Murphy's law!

Chapter 11

Communicating Safety in Construction

Introduction

The health and safety of people has to be the highest priority of every individual or organisation in any activity, no matter what business or industry sector is concerned. As individuals we all endeavour to safeguard ourselves and others from accidents and injury. Moreover, whether an employee or a member of the public, every person involved directly or indirectly with an organisation or its activities has to be protected against accidents, prevented from causing accidents to others and educated in identifying potential accident situations. This is an ethical, moral and essential obligation on us all. Coincidentally, safety and safety issues are also fundamental to business success.

Nowhere are these sentiments more vividly illustrated than in the construction industry, which has a reputation for its inherently dangerous environment and susceptibility to accidents. Uniquely, the industry is outdoor and open to the elements, but at the same time it is dependent upon the use of large and heavy mobile plant, a wide variety of energy-intensive equipment and, unlike farming, employs a diverse range of multidisciplinary professions and crafts. The nature of the business also makes the work environment open to the ingress of non-core personnel, including subcontractors, suppliers and the public at large. In all, it is a pot pourri of dangerous situations.

Specific hazards in construction

The hazards associated with construction are well documented. The main areas of concern as identified specifically for site safety are as follows:

- Excavation
- Demolition
- Work at height
- Buried services

- Mobile plant
- Work in confined spaces
- Work over water
- Manual handling and portable tools
- Diving operations
- Exposure to hazardous substances, dust and fumes
- Contaminated land
- Temporary works
- Fire
- Electricity
- Noise
- Vibration
- Piling
- Roadworks.

All of these areas of concern affect employees, subcontractors, suppliers and anyone else connected with operations on the site. But they also affect, and therefore present a risk of injury to, any unauthorised persons on the site. Security and boundary fencing cannot guarantee protection from unauthorised entry. Many of these areas could also affect others off-site. The general public can be severely affected by noise, dust, vibration, increased vehicular traffic and other elements.

A partnership between the safety department and public relations

Safety is clearly a major contributor to the reputation of a company, and is the major concern of the safety department of a construction organisation. Reputation is the major concern of public relations. This chapter aims to demonstrate how a practical partnership between the two departments will not only bring mutual benefit, but will provide immeasurable benefits to the whole organisation.

What do we need to do?

The objectives for the public relations function within the area of safety are manifold. Accidents cost lives and are bad news. Injuries cost time in man-hours and productivity. Accidents affect families, relatives, friends and colleagues. All affect the name and reputation of a company. The public relations practitioner has to reinforce the name and reputation of an organisation, as well as to communicate with its employees.

The main objective, therefore, is to create, protect and enhance a safety culture throughout the organisation. This means **communicating safety** and helping to educate, promote, publicise and reinforce safety issues wherever appropriate *in support of the safety department*, to every audience that can be reached where it is essential or desirable.

The audience is, first and foremost, the employees. From the point of the accident to the managing director, accountability is clear. Therefore, from the top to the bottom of an organisation, basic safety messages have to be communicated. The safety department is obliged to communicate the legal, technical requirements through its own established channels and those specified by the rules and regulations dictated by the Health and Safety Executive (HSE) and other authorities. However, the public relations function should not only reinforce those messages, but also promulgate them further in simplified, layman's terms to promote and reinforce a safety culture on a broader basis than that reached by the official safety outlets.

If these objectives are met, they will in themselves help reach further safety/public relations objectives. Employees need to be aware of direct, professional or industry-related safety issues. If this is achieved, they will also become aware of the wider safety issues. In this way, their attitude should reflect on their dealing with subcontractors, suppliers, and anyone else visiting a site.

Site safety, beyond the immediate safety issues applying to all activities (welding, scaffolding, toe-boards, helmets, reflective vests, etc.) must also include all personnel affected by those works. This inevitably includes off-site people, usually the public at large. Here, site-based safety awareness is crucial if attitudes are to be communicated to the onlooking public outside the site gates.

Public relations not only includes the range of site safety posters that adorn the outer gates and fencing surrounding a construction site. It includes dust, noise, mud and water pollution from a site, all potentially dangerous elements that are carried outside the perimeter of a site. Even these considerations need the close scrutiny of the public relations practitioner.

This leads on to a further objective: the education of the external audience. On the surface, a tidy site, with clear, organised safety instructions and posters, reinforced by obvious and tangible adherence by site personnel to safety issues (i.e. *all* wearing helmets all the time, and *all* wearing high visibility clothing, etc.) will certainly impress the onlooking public. They will be made aware, by the image given over, that safety is important, indeed essential. However, this must be reinforced if lives are to be saved and accidents avoided among the public in and around sites.

A key objective of the public relations function, therefore, is to

press warning people of specific safety issues. For example, people must keep off the beaches when heavy rock armour is being landed and handled by track-mounted, grab- or bucket-equipped excavators placing rock in a coastal protection scheme. It also includes active liaison with schools to educate children in safety awareness.

All of the objectives of the public relations function in support of the safety aspects of the construction industry are linked inextricably and intricately to the legal obligations of the industry to the Health and Safety Regulations. Nothing the public relations function does in support of promulgating safety messages must trivialise. Nor must it patronise, nor over-simplify. Every single safety issue is very serious and must be handled soberly, rationally and especially sympathetically to the end-user – the human being.

Safety law and communication

In this context, the safety issues are extremely complex. Beyond the basic human and practical aspects, there is a vast array of legal obligations, all of which need communicating. These range from all-encompassing Health and Safety Executive rules and regulations backed by the full authority of the law through a variety of Acts of Parliament, under the umbrella of the Health and Safety at Work etc. Act 1974, to directives from the European Union Commissions in Brussels concerned with safety issues. This is continuously reinforced by regular reviews and regulations being published by the Health and Safety Commission (HSC).

It is not in the scope of this book to delve into the intricacies of the health and safety laws affecting the industry. It is a legal requirement that all construction engineers be aware of their legal obligations. However, the public relations practitioner needs to be aware of the main issues and developments affecting the industry's safety procedures and regulations: *communicating* safety makes the greatest contribution of all to improving safety records and reducing accidents.

The Health and Safety at Work etc. Act 1974 was aimed at securing the health, safety and welfare of people at work, and also the protection of other people not at work, against risks to their health or safety arising out of or in connection with working activities. A company or an individual may be charged and convicted by a criminal court if they have been found guilty of contravening the Act. This Act also set up the national safety agency, the Health and Safety Executive (HSE) through which rules and regulations would be implemented and supervised.

All employees, no matter what industry, are particularly obliged through statutory duty to comply with sections 7 and 8 of the Health and Safety at Work etc. Act 1974. This includes taking reasonable care of the safety of themselves and of any other persons who may be affected by what they do or fail to do at work. All employees, under the same Act, have to co-operate with their employers or any other persons in the performance of their statutory duties for safety matters. Similarly, they must not misuse or interfere with anything provided in the interests of health, safety or welfare.

In addition to general health and safety laws as defined above, there is also a myriad of legislation and directives specific to the construction industry itself. The latest catch-all is the Construction (Health, Safety and Welfare) Regulations 1996, which came into force on 2 September 1996. These regulations implement certain provisions of the temporary or mobile construction sites Directive, and replace three sets of existing UK legislation. They also contain many new requirements such as precautions against falls, and traffic routes to separate vehicles and pedestrians. The public relations practitioners should be aware of the broad issues raised by these regulations.

In a swiftly changing industry, one of the most influential pieces of recent health and safety legislation to affect the construction workplace was the Construction (Design and Management) Regulations (CDM), which came into force on 31 March 1995. Where Design and Build (D&B), Design, Build, Finance and Operate (DBFO), and Private Finance Initiatives (PFI) have become a major part of the construction industry, CDM Regulations have had a profound impact on all construction employees. Communication has become a paramount aspect of these regulations.

The regulations apply to all sites that employ more than five people and to those that last longer than 30 days. In essence, CDM imposes new duties upon clients, designers and contractors for the establishment, implementation and monitoring of all safety issues in construction. In so doing, the regulations place more responsibility on clients and design teams rather than just the contractor. Clients have to ensure that the construction phase of any project does not start unless a comprehensive health and safety plan has been prepared. They also have to appoint a planning supervisor to oversee all safety aspects of the project, including safety elements in the design, and to liaise with HSE on project notification.

The contractor still has full responsibilities for health and safety plans and their management, and for safety training. However, the designer, whether as consulting engineer or as contractor (in Design and Construct contracts), has to 'ensure the design, if possible, completely avoids risks to

Of vital importance under CDM is the need for a construction company to be able to demonstrate and prove to a client or potential client that it has an efficient safety policy and systems in place to ensure an acceptable safety standard. A company's safety performance is regularly scrutinised by clients who have a duty under CDM Regulations to appoint a principal contractor that is 'competent' and provides 'adequate resources' for health and safety matters.

CDM, in tandem with all the other laws, rules and regulations and backed up by individual company or group-based rules over and above the legislative situation, has given construction organisations an even greater need to get the safety message over to every employee – by both carrot and stick methods. In this context, the most important elements of any safety policy within the construction industry are communication, training and management. In turn, training and management also depend upon communication.

The most important communications must, by definition, be those at site level, initiated, supported and reinforced by top management and directed by safety officers. Any reputable company will have a very strong communications programme directed by its safety department and safety officers through area/divisional managers, contracts managers, agents, section engineers and foremen to individual plant operators, gangers and operatives. It is a cascade of information.

This cascade, in safety terms, has to be planned, managed and monitored for effectiveness. It has to be two-way, hence the need for reporting systems. Again, every construction company is obliged to have a safety policy system established that ensures immediate and comprehensive reporting and feedback of safety information. The foremost written statement is the safety policy document, and perhaps the chairman's statement. Some larger companies develop their own health and safety procedures.

Where does public relations fit in?

The public relations function *should be* a major element of the safety policy. If communications are so fundamental to the sector, then the public relations practitioner must be involved integrally with the safety department, not only as a partner in promulgating key messages, but also as an advisor and 'consultant' in assisting the safety officers to give out and reinforce their own, more technical, messages in a professional, understandable and enforceable form. Presentation of the message can be as important as the message itself. If people won't listen, the message will not be received. It is as simple as that.

The main elements of support that the public relations practitioner can lend to safety can be categorised as internal (employee communications, including 'cascade flow' to subcontractors and suppliers and, indirectly, to client and engineer) and external (pre-qualification evidence to clients and engineers, the public, the press and media, the regulatory authorities and potential clients).

The range of public relations communications must also be categorised as:

- Preventative information (educational)
- Legislative (legal obligations)
- Reportive (feedback and the quality circle)
- Cultural (staff awareness and an ingrained culture for safety situations)
- Obligatory (spelling out specific dos and don'ts).

In summary, the public relations function should be one of a number of cornerstones to a company's safety policy by providing some or all of a range of services, some of which will be covered in more detail later in this chapter. These services are as follows:

- *Advising, assisting and educating* the safety department and officers about all presentation and communications media.
- *Safety newsletters* distributed to every area office and site giving topical information on practical issues and the latest legal/regulatory matters.
- *Safety posters* – above and beyond official HSE-issued and commercially produced posters. Mainly incident specific or of a general, very basic nature to reinforce obvious safety precautions and regulations.
- *Staff newspapers.* In any company staff newspaper, safety issues must be raised, profiled and highlighted and messages reinforced – especially when the staff newspaper is distributed to employees' homes, where family pressures will reinforce safety messages, particularly basic ones such as the need for head protection, gloves and toe protection.
- *External marketing communications.* Any publicity brochure or leaflet which aims to highlight an organisation's skills in general construction, civil engineering, or specific sectors must not exclude safety elements. Even a leaflet on highway maintenance needs to contain reference to a company's safety objectives, standards or systems. If brochures are to be used to convince clients or potential clients of a company's capabilities in any sector of the construction business,

safety in this context.

- *Internal management information systems.* These include noticeboard messages, staff circulars, management meetings, board meetings and any other medium through which the management gives out messages and information to staff and receives feedback from employees. This could include the annual discussions between directors and staff relating to annual turnover/profit figures and objectives for the coming year. Every occasion should include safety issues in a prominent place within the overall information flow. The 'Investors in People' programme demands that we communicate with all staff. Use this to communicate safety.

- *Press and media.* Safety information needs the widest circulation. The media offer the opportunity to promulgate safety issues more widely than most other methods. They should not be used casually. But they can be used very effectively to help specific site-safety issues where public safety could be at risk. There are times when only a close liaison with the local newspaper, radio and television stations can help prevent the ingress of the local population, especially children and youths, from jeopardising their own safety in or around a construction site.

- *School liaison.* A construction site can be a magnet to adventurous children. Adventurous children in the proximity of deep excavations, pipes, heavy plant, equipment and other materials spell danger. No matter how many posters are displayed around the site perimeter, they will not dissuade the adventurous child or youth from entering the area, whether to play or to vandalise.

    Site security is one answer. Community relations is another. As soon as a site is established, it is good public relations for every school in the area to be contacted and offered a safety talk by a company safety officer accompanied by an engineer who can talk about the specific dangers of that particular site. School talks also tend to move the messages back to parents within the community and therefore back to the general public around a construction site. This, in its own right, is good public relations. Moreover, it helps keep children away from construction sites.

All of the above listed activities undertaken by the public relations function are there to reinforce the main thrust of the safety department's own information communication system. They do not in any way eliminate the need for *operational* safety information communications.

Communicating to your own people about safety

This section covers three methods for communicating safety:

(1)   Operational safety communications
(2)   Safety newsletters, posters and campaigns
(3)   Staff newspapers.

Operational safety communications

In this context, the public relations function can help with the best methods and skills to promote safety within structural, managed-cascade information systems. In most construction companies, the main methods of communicating and reinforcing safety issues to staff and operatives are part of the safety policy structure.

A company safety policy will be based upon a safety manual which contains comprehensive information on best practice for health and safety in construction and detailed references to relevant legislation, codes of practice and guidance. This will be regularly updated and issued in a controlled manner to managers. Within this manual, communication is a major feature, whether part of the establishment of method statements, training, monitoring, auditing or accident recording, reporting and investigation.

Before any project gets under way on site, an initial risk assessment is undertaken under a health and safety plan. This is then communicated to all operational managers. The pre-start risk assessment may identify certain activities which require a method statement. This is a comprehensive step-by-step account of how the works are to be executed safely, and is used to identify:

- potential hazards to persons on site or the general public;
- risk of damage to plant and structures;
- difficulties which may be encountered in carrying out the work;
- special plant or procedures needed;
- how the work is to be carried out safely;
- liaison with others who may be affected;
- permits or licences required;
- how the procedures will be communicated.

This mechanism and the site safety plan are designed to ensure that all the necessary safety information about the site or operations on it can be communicated to all parties. The methods utilised for the promul-

the following:

Accident reporting, recording and investigation

If accidents are to be prevented, we must learn from past experience. Unless accidents or even 'near misses' are reported, lessons will never be learned by others in an organisation.

The responsibilities for recording, reporting and investigating accidents generally rest with line management, assisted by the safety department. The systems employed are designed to:

- satisfy statutory reporting requirements;
- provide measures of safety performance;
- provide information on areas of activity which require attention.

This latter point in particular enables the safety department to use reports as the basis of company bulletins to highlight particular incidents that, by planning, can be avoided or eliminated on other sites.

Tool-box safety induction talks

These are generally given by the foreman or works manager or, on a small site, the agent, to all new operatives employed on a site and also on a regular basis to reinforce issues to all operatives. The subjects are usually of a general construction safety nature, backed up by specific safety situations pertaining to that particular site. For example, if water (river or sea) is an element of the works, then buoyancy harnesses, life-jackets, etc., will be prioritised as well as basic helmets, high-visibility clothing and the like.

Tool-box safety induction talks must not be treated as a lecture only: they need to be an opportunity for discussion, to raise awareness of any problem areas, to suggest improvements, and to share experiences from other sites. Open talk about safety issues needs to be encouraged.

Leaflets, booklets, handbooks

These include the general health and safety policy handbooks issued by most companies which give every single employee the full range of safety rules and regulations covered by the company as well as the legal obligations placed upon every employee. However, they also involve simple safety 'cards' issued to any plant operator or labourer involved in specific tasks. These range from the rules and regulations concerning banksmen, rules to drivers entering a site, the use of shops for craneage and lifting situations, to such basic safety issues as the use of abrasive machinery.

Training programmes

These apply particularly to formal training for managers within construction, and are to be 'cascaded' by managers to sites and reinforced by tool-box talks. Comprehensive training courses are given to all concerned, from directors and senior management to site agents and foremen, on what regulations involve and how they are to be managed.

The 'Investors in People' safety scheme

A company committed to the Investors in People scheme needs to communicate with and develop all employees. In this context, safety training and safety awareness have an even higher priority throughout a company. All construction operatives, craftsmen and tradesmen will, at an appropriate stage, have to be given a safety awareness course by safety officers within a formalised, phased programme.

Based upon CITB guidelines, the aims of the overall Construction Skills Certification Scheme (CSCS) are to improve health and safety awareness, to recognise skills, competence and qualifications and to identify training needs to improve or add to existing skills. The safety training element covers the full range of occupations within construction and civil engineering works, and courses lead to a certificate of completion.

HSE warning and safety signage

The HSE issues a range of signage regulations concerning every aspect of site safety, from basic head protection to specific site situations relating to water and noise hazards, deep excavations and contaminated materials. Companies are encouraged to customise their own signage in tandem with legal requirements concerning safety signages to reinforce specific safety messages to staff and the public.

HSE videos/publications

The HSE has produced a range of safety videos and publications/leaflets, not only of a general nature, but also for specific safety issues concerning certain conditions and situations, e.g. contaminated land operations.

Safety competitions

These can be in-house safety competitions between sites or areas, or they can take the form of participation in the HSE or Institution of Civil Engineers (or others) safety competitions held throughout the country on a regional and national basis. All encourage a greater awareness and knowledge of safety rules, regulations and legal obligations.

All of these elements are simple supplementary activities of the formal site-based safety procedures and safety officer inspections. None replaces the official, hard-nosed safety programme and systems implemented through law on every construction site. However, communication remains a vital part of each of the elements in assisting safety officers to promulgate their messages.

Partnership needs to be established between the safety and public relations departments. When safety officers have their monthly or quarterly management meetings, there should be a slot once or twice a year for the public relations manager to attend, to listen to the latest key objectives and messages that need disseminating and to contribute ideas on the best methods for giving out such information. Likewise, the public relations manager needs to keep informing the chief safety officer about various communications routes to explore within an organisation: new ideas, media, exhibition events, and training systems, for example.

The partnership between the public relations and safety departments also extends to the mutually supportive and beneficial elements of both functions where overlap is essential. As will be illustrated later in this chapter, staff newsletters/newspapers, press and media management and schools liaison are all areas in which overlap is particularly evident. They are the main tools for promulgating the safety message, both internally and externally.

Safety newsletters, posters and campaigns

Most construction companies of calibre will have a newspaper or newsletter published on a regular basis and dedicated exclusively to safety issues. Although these are published by the safety department of an organisation, the public relations function should have a major role in writing, designing and producing such newsletters. As professional communicators, they may be better equipped as writers and designers than safety officers. In partnership, therefore, a better, more professional safety bulletin or newsletter can be produced, giving greater levels of readership and understanding.

A safety newsletter has to be just that: topical news about real safety issues that affect people undertaking their everyday employment activities. But it also must be as simple as possible if its main task is to communicate issues to the site-based operative, labourer, plant operator and so on. The messages need to be simple and concisely worded, and the headlines large and clearly targeted on the subject.

One of the objectives of the newsletter is to inform employees regularly about new health and safety regulations being issued by the HSE that affect all of them as they undertake their work. This is one of the most difficult aspects to put over simply and in an interesting way so as to guarantee readership. However, it is essential that this information be successfully promulgated. And it is equally essential that the message be totally practical and aimed at real, understandable operations that can be recognised by all site operatives.

For example, the new Construction (Health, Safety and Welfare) Regulations 1996 which came into force in September of that year were a complex compilation of rules that implemented certain provisions of the HSE Directive on temporary or mobile construction sites and replaced three sets of existing UK legislation. Within the scope of a newsletter circulated to every site for placing on noticeboards in cabins and canteens, an explanation of these regulations would have been a nightmare: complex, complicated, legalistic and everything that defies simple interpretation and communication to operatives.

However, it was essential that the main elements of these regulations as they affected everyday operations were highlighted in a newsletter format that could then be backed up by more detailed information given out by site managers and then foremen. In this context some companies opted, quite rightly, to select just a few prominent, salient points that would have immediate impact upon most site-based personnel.

One company issued its safety newsletter in August 1996 with the banner headline:

'New regulations on 2nd September!'

In two sentences the lead article introduced the new regulations, with the main thrust being: 'They contain many *new* requirements, including...'. Then, in very simple terms, it picked out two of the most obvious elements that would affect most employees, and it put the details in the simplest and most practical words:

The main guard rail must be at least 910 mm high and toeboards a minimum of 150 mm high. There must not be an unprotected gap exceeding 470 mm between any guard rail or toeboard. This means that an intermediate rail will now be required, or some other means to prevent any persons or material from falling.

Traffic routes. All sites must be organised so that pedestrians and vehicles can both move safely and without risks to health.'

To prove that the message was extremely serious and that further back-up information would be forthcoming, the same article concluded with a very short, sharp warning about how important the regulations were, how soon they would be implemented and how further information would be forthcoming:

'There will be no lead-in period with these Regulations which will be **fully in force as from 2 September, 1996.** A safety bulletin will be circulated giving more details and the regulations will be an integral part of all future health and safety training.'

A newsletter must also be the means of reminding staff about existing regulations, especially those covering situations where a company has identified a weakness in implementation where incidents have occurred within the company or where the industry itself has been the subject of a high-profile accident affecting a particular aspect of work practice.

Some company safety newsletters have a regular, boxed, short piece every issue reminding employees of a given regulation or even reminding them of simple precautions against injury. These again need to be very short and very simple and must always relate directly to practical advice. Good examples of these reminders follow.

**A Reminder: Chapter 8, Traffic Signs Manual
Sections 65 and 124, New Road and Street Works Act 1991**

High visibility warning vests/jackets
BS 6629: 1985 has been replaced (September 1994) by BS EN 471: 1994 High-visibility warning clothing. **Failure to comply is a criminal offence.**

All personnel working on or near a carriageway must be readily visible to all drivers. To this end, high visibility garments must be worn with at least two 50 mm bands and braces of retroreflective material. Such vests are available to all personnel – get them and wear them: if you don't you're breaking the law. Moreover, in the summer sunshine these vests may help to prevent skin cancer.

A Reminder: Regulation 34(1) Chains' Ropes and Lifting Gear
The Construction (Lifting Operation) Regulations, 1961

No chain, rope or lifting gear shall be used in raising or lowering or as a means of suspension unless:

(a) it is of good construction, adequate strength and suitable quality; and
(b) it has been tested and examined by a competent person and there has been obtained ... a certificate of such test and examination; and
(c) it is marked ... with the safe working load and means of identification.

Employees can be assured that lifting gear officially provided meets the above criteria by checking the current colour code – the correct colour code shows that the equipment has been recently inspected and certified for use.

A Reminder: Regulation 21, Construction General Provisions Regulations, 1961

(1) Wherever work is carried out in enclosed or confined places (pits, shafts, tunnels, etc.) where the air may become deficient in oxygen or contaminated by dangerous dust, fumes or gases (e.g. from explosives, exhaust gases, etc.), adequate ventilation must be provided to ensure the atmosphere is fit to breathe.
(2) If the air in any enclosed or confined place is suspected of being poisonous or asphyxiating, tests must be made and no one allowed entry until the place is clear.

Any confined space on your site has been or will be identified and a permit is required prior to entry by any personnel. In any confined space you can be overcome by fumes or lack of oxygen without warning; as can any rescuer. You **do not** enter such a space unless a permit has been given and you must have received safety training specific to these conditions.

Topicality is also critical to the success of a safety newsletter. The topic can often be seasonally based. As the summer school holidays approach, for example, a safety newsletter can home in on safety related to children:

'As school summer holidays get into full swing, all site staff must redouble their vigilance in keeping children off our construction sites ... For those needing more information on how to keep children off sites and duly warned, the Health and Safety Executive publishes Guidance Note GS7, *Accidents to children on construction sites*, ... copies are available from the Safety Department.'

newsletter was given over entirely to this topic. It started off simply stating:

> Most of us in this industry spend a great deal of our time outdoors. As we go into the summer months it is essential that we remember the health risks from working in the sun. This issue reminds you to cover up – not only with a helmet, but also with a shirt and reflective vest.

It then continued by giving a series of bullet-point, simple answers to the questions 'What are the dangers?', 'Who is at risk of skin cancer?', 'Can I protect myself?', and so on.

Other topical features can be linked directly to specific incidents affecting the company itself, i.e. a recent, real accident, or incidents reported in the construction press and given a high profile. That way the story can be given a genuine news basis that will be of interest to the reader. For example, one way to highlight concern about dermatitis amongst construction staff was to refer to the technical press:

> 'A recent issue of *Construction News* highlighted the Health and Safety Executive's (HSE's) real and proven concern about the continued scourge of the crippling skin complaint dermatitis amongst construction workers.
>
> Although the article referred primarily to tunnelling workers on the Jubilee Line Extension, it spoke of the HSE's new campaign to stamp out occupational dermatitis. It is estimated that more than 10 000 construction workers are suffering severely from the disease.
>
> We must all be aware of the dangers of dermatitis, and supervisors must spot any tell-tale signs of the disease developing in their workforce. Inspect your hands – if you are concerned, consult the first aid representative on your site.'

All of the messages or issues targeted by a safety newsletter can be strongly reinforced with suitable posters. Together, a newsletter and accompanying poster can be the basis of safety campaigns that the safety department identifies as topical and essential. The HSE issues many posters and there are many that are available commercially. However, there is no substitute for the customised poster produced in response to a specific issue or type of activity.

Like the newsletter itself, the accompanying posters need to be simple and easily reproduced. Issued in A4 format alongside or with the newsletter, these posters should be easy to photocopy and to enlarge on a photocopier to A3 for display in cabins and canteen walls as well as other areas where numbers of operatives assemble.

Safety posters can range from simple reminders to employees to keep their helmets on 'even from cab to canteen', or they can be produced to give a safety alert message: 'Before moving anywhere close to an excavator or crane that is operating, *alert the driver* so he is aware of you!' Regular posters reminding operatives of equipment availability are also effective, whether it is for respiratory protection, gloves, goggles or ear protection.

All of these topics that are fundamental to a successful, dedicated safety newsletter can also be carried over into a company's ordinary staff newspaper. This double exposure does nothing to detract from the messages themselves. They will only be reinforced.

Staff newspapers

Staff newspapers are often the only medium through which a company and its senior management can communicate with every member of staff. Noticeboard notices, circulars, reports and so on will often allow much of the intended audience to slip through the net. A staff newspaper usually has a broad and comprehensive 'hit rate' as it generally covers people and social events that are of interest to most employees. If the newspaper is posted to all employees' homes, then the impact is even greater. Thus, the staff newspaper is an invaluable tool in the promotion of safety and safety awareness among staff.

Whether the publication be weekly, monthly or even quarterly, safety needs to have a degree of prominence within its pages. Some companies have a safety column specifically assigned to each issue of their staff newspaper. Unfortunately, this use of a regular, identifiable spot or column within a staff publication can be counter-productive. It is expected and anticipated, and therefore is seen as predictable, authoritarian, dictatorial or even 'big brother' orientated in the eyes of the reader.

The safety column approach should not be dismissed as it gives due and prominent recognition of the topic. However, familiarity can often breed contempt or contentment, neither of which should be associated with safety.

It is preferable and, arguably, more effective, to present safety topics within a staff newspaper as genuine, stand-alone articles in their own right. If any safety issue is not interesting or topical, then it will not be read. It is the same with any story in any newspaper. Unless the heading and/or photograph is interesting, the article will not be read. The 'Safety bulletin' or 'Safety is important' type of headline in a newspaper will

A public relations practitioner will usually take responsibility for the contents and presentation of the staff newspaper. As such he or she has a direct link between the safety department and all staff. This is an invaluable conduit for the flow of extremely important information. It represents another critical element in the chemistry between the safety and public relations departments that is so important to a company's success.

Whether annually, quarterly or in every issue, the chief executive, managing director, chairman or president of an organisation will have an opportunity to make keynote addresses to every employee through the staff newspaper. This is where the public relations practitioner needs to make sure safety is raised as a cornerstone issue by the most influential management within an organisation.

This could coincide with the annual report and accounts issue, the half-yearly results issue or the Christmas message to staff, or form part of the announcement of a new corporate acquisition. Whatever the main reason for the senior executive to publish his own words within the staff news-paper, safety must be built in somehow. It will be read alongside other obviously important messages.

One major European construction company gives four occasions a year for its president of the board of management to address its nearly 5500 staff throughout Europe and overseas within the individual staff news-papers of its 15 group companies. These are a Christmas message in December, the annual results in April, a summer message in June/July and half-year results in September/October, irrespective of how fre-quently each staff newspaper of the operating companies is published. Without exception, every single message or article from the president contains an element dedicated to safety. The latest message includes an effective reference to the issue, put in a personal and understandable way:

'... To constantly strive for improvement is no easy task, but only by doing so can we safeguard the future of our Group.

Likewise, it is only through constant effort that we can make our construction sites safer places to work. Again this year, [Company/Group] employees – your colleagues and mine – have unfortunately been involved in accidents. We need to be alert to our own safety and the safety of those around us at all times. Safety awareness must remain at the core of our efforts to achieve continuous improvement...'

There should also be an opportunity for the chief safety officer to contribute an article to the staff newspaper at some stage during the year. This need not be a regular feature but a one-off article related to a specific issue, or announcing a new issue introduced into the working life of employees, new responsibilities or new legal requirements affecting all staff.

Luckily, a staff newspaper has access to numerous angles from which to cover safety issues. The angle could be a personal, individual action or an activity that is newsworthy; it could be an event such as a safety quiz; or it could be of a topical nature such as sunburn threats during the summer months. In each case, the word 'safety' might be omitted as an irrelevance in the title but is of fundamental importance within the context.

Good practical advice is usually the best way of communicating with construction staff. And if this means repeating messages already promulgated in dedicated safety newsletters (discussed in the previous section of this chapter) then it matters not, so long as it is both topical and can be associated directly with real-life site-based operations. Illustrating this is the following article carried in a recent issue of a staff newspaper of a civil engineering company:

'Look before you leap'

Despite a good safety record by industry standards, too many employees are involved in accidents on construction sites in the UK.

Most of the injuries are at least partly due to the non-observance of basic safety precautions. In short they are avoidable.

A significant proportion of reported accidents concern 'struck by...' incidents. Therefore, over recent months, all safety officers in the company have been driving home the message about taking precautions to avoid being struck by plant and vehicles on site.

This has been supported with four themed posters in the bi-monthly publication, *Safety Scene*. The posters, which are best blown up to a larger A3 size, carry strong cautionary messages:

Be seen! High visibility clothing saves lives

High visibility clothing is a good idea on any site but it is required by law in roadworks, rail sites, and on sites where there is heavy traffic movement. Failure to do so could result in prosecution, injury or worse...

Keep clear of swing. Can the driver see you?

Vehicles with a backswing like cranes or excavators represent a major danger. The procedure is always to alert the driver to your presence before walking behind a vehicle. It is also good practice for site managers to avoid situations where vehicles are required to reverse, if at all possible.

The Construction (Health, Safety and Welfare) Regulations contain details about how sites must be organised. Pedestrians and vehicles should be kept apart wherever possible – with separate site access points and exits, and barriered footways.

Keep clear of ALL machinery

The use of trained banksmen to control high risk situations is recommended. Temporary roads should be prepared with proper running surfaces to avoid skidding. Steep gradients, cambers and sharp bends should also be avoided when laying out a site. All represent potential dangers to personnel working on a site.

Any excuse for an article on safety should be used for the staff newspaper. It is the prime objective of every construction company to bring to all of its staff a safety culture, a culture that has safety in-bred to every activity an employee undertakes. Only by constant and continuous coverage of the issues through safety newsletters and staff newspapers can it become integral to each person's thought processes when they undertake their everyday activities.

To bring all these elements of internal safety communications media and methods together requires a powerful array of tools that the public relations function can provide to the safety department of a construction company. The tools are all essential to meet the overall strategy of any safety improvement campaign.

The chief safety officer of one of Europe's largest construction companies recently described the development of a safety programme as having four distinctive levels:

'It starts with the registration phase. A policy declaration is made, accidents are categorised and the matter is placed regularly on the agenda. Then we start to organise safety. Procedures are set up and inspections introduced. The third phase is oriented more towards prevention. Plans of action are evaluated and we set ourselves new goals for improvement. Finally we must achieve a position where safety is fully integrated into every level of the organisation and has become completely natural to everyone. That is where we have to go.'

Press and media

The influence of the press (newspapers, technical journals, etc.) and other media (television and radio) upon the image of any organisation is profound, in both the positive and negative senses. The media have an equally profound effect upon the perceived safety consciousness of an organisation and can either create or destroy reputations in the context of safety. It is wrong to say one can 'manage' the media in the context of safety, because that implies the need to cover up or to create an image. However, sensible, practical public relations activities and actions can reinforce a reputation for safety where one already exists.

More importantly, the public relations function can use media outlets as a remarkably powerful influence to *increase* safety on a construction site. It can also use them very effectively to help drive home extremely important, even crucial, issues about safety to colleagues not only within an organisation but, perhaps most importantly, throughout the construction industry. The media are a major tool in the range of available conduits for safety information messages and, used wisely and pragmatically, can become the main cornerstone for building up the industry's somewhat tarnished reputation in the field of safety.

Hiding bad practice, covering up potentially hazardous situations, pretending to have effective control over a dangerous operation may provide short-term protection from a vigilant journalist covering a site's operations. However without an immediate and very strong internal reaction to those incidents reported not only to the safety officer but also to the site managers and, if sufficiently compromising to the company's reputation for safety, to the main board director responsible for safety, *the lessons will not be learned*.

The public relations manager, in this context, has to be the eyes and ears of the company's outer image for safety and for ensuring its genuine commitment to protecting its workforce and the public against accidents or injury. No matter how unpopular it makes him or her, the public relations practitioner has an obligation to highlight any areas of a site that may lead to accusations that the safety of personnel or visitors is being compromised. To ignore such situations is to ask for ongoing problems in the next site scenario and does nothing to help avoid a potentially disastrous press and media crisis should an accident or near miss be reported.

Among the most influential readers of the technical press is the client's engineer. If he or she sees a company name associated with legal proceedings following a site accident, guilty verdicts, fines, etc., then instantly the alarm bells will ring next time that contractor submits its prequalification documents for a contract. If only subconsciously, a negative safety incident plays an enormous role in the decision-making mind or process of a potential client.

There are three distinctive aspects of the press and broadcasting media in the context of safety within the construction industry. One is the

image of an organisation; and the third is the proactive use of the media to enforce or reinforce a safety situation, both at site level and at corporate, company level.

Emergencies or incidents involving safety or accidents usually and quite quickly escalate into crisis situations. It is therefore logical to discuss here the management of accident situations and the role of the press and other media.

Emergencies, incidents, accidents

Critical to both the safety and public relations departments and of greatest potential impact upon an organisation is the subject of incident or crisis management, and other site emergencies. Every construction organisation intends to ensure that any risks arising from its work activities are either eliminated or reduced to a minimum. It is acknowledged by most, however, that, despite these measures, there is always the possibility of a major incident. The broad issue of incident management is considered in some depth in Chapter 12. However, it is also inseparable from the safety issue.

The safety department and each site make arrangements for handling the more likely emergency situations, i.e. personal injury, fire, work over water and the like. This includes all the practical measures of liaison with the emergency services, first aid provisions and investigations.

These measures are often reinforced or added to by outside parties. Where work takes place on sites for which emergency procedures are already in place (e.g. Control of Industrial MAjor Hazards (CIMAH) sites such as chemical plants) or sites for which emergency procedures are dictated by the client (such as Railtrack and London Underground Limited), construction managers must ensure that such additional procedures are fully implemented.

Within all such scenarios, the public relations function needs to be interactive, not just through media management but also through internal and client/public communication. In particular an *incident management plan* should be drawn up specifically in relationship to the communications network for each site. This is an operational, working document that specifies the responsible persons and contact details for incident-reporting lines.

Ultimately, the main role of the public relations manager in the emergency or incident management situation is to control and manage the external communications emanating from the situation. Although this may well involve liaison with nearby communities that might be affected hy an incident, the multi priority will be media information. This takes us on to press and broadcasting media relations as part of the safety communications package.

The way in which an incident is reported is also of critical importance to the tone of the resulting article or feature broadcast through the media. From where did it first emanate? Who was interviewed? Who was referred to? Which names were given? What supposed 'facts' were talked about? In this context, most of the procedures are dealt with in Chapter 12, which covers incident management and emergency issues. However, safety by its very nature overlaps with crisis management. Most 'crises' in this section involve safety issues, either potential or actual. Therefore general incident management is interlocked with safety implications.

When a journalist, whether local, national or technical, rings up the public relations manager asking questions about an accident on one of your sites, the chances are that the original source was the police or ambulance service. It is vitally important that the public relations department be informed of any such situation *before* any external party (except the emergency services!) and parallel to the safety department and directors.

The journalist has often had the most dramatic description of a situation from the emergency services that may totally distort a situation if it is not put in the context of a construction site and its operations. If the press officer or person responsible is fully aware of a situation or occurrence and knows the full facts, then a very quick practical response can calm a situation. However, quick off-the-cuff and reactionary responses that have not been verified with the necessary authorities can plunge an organisation into greater depths of crisis. That is why there must be an incident management plan.

It is important that no immediate, unprepared comments be issued by any organisation off-the-cuff to a given situation. The incident needs to be thought about fully, and the reaction to any statement needs to be considered. Moreover, all parties concerned must be borne in mind. Is your client impacted by either the incident or your response? Is the consulting engineer's position compromised in your response to an accident? So many questions can arise, and an instant, unplanned response can give very confused messages, implying far more than is required and often compromising the situation and stance of other associated parties.

As discussed in Chapter 12, any incident management plan needs a readily available, multipurpose statement to give out to the press in an instant response situation. This is most important for the site agent, and particularly for the safety officer who may be the first 'expert' on the scene. The safety officer might be extremely professional about safety issues, but will surely be unqualified to deal with urgent, hard and

Within any plan, therefore, the communication network has to be firmly established and adhered to. Only one spokesperson should be permitted and all information needs funnelling directly to that person. The spokesperson also needs to control how that information is released: there are many audiences out there begging for information in a variety of forms, from technical, medical and emotional to financial, factual and hypothetical. No glib answers or off-the-cuff remarks can be given without real assessment.

Most journalists, whether technical, local or national, will understandably report accidents or safety issues in purely human terms. However, emotive reporting of an accident usually distorts the reality of a situation. Inevitably the seriousness of very minor incidents may become greatly exaggerated.

Any response to the press about any incident must therefore pull the angle back from the emotive to the practical. An accident usually results from human error, mechanical failure or very explicable events leading up to an incident. The emotive must quickly be brought back to the rational by trying to identify the mechanics of what happened. Thus a bold, practical response that is divorced from the specific incident concerned, but that addresses the basic issues, is essential.

In any event, the response has to be concern for the party or parties affected; reassurance that a full investigation is being undertaken; commitment to the application of resources and specialists to fully assess the incident and fully report lessons to be learnt; and, finally, that every outside party, from emergency services to Health and Safety Executive officials and from local authorities to clients have been fully appraised and consulted. Further information must also be given as and when factual details are available.

The recommended response document(s) is shown in Chapter 12. The most important element of this is the referral of any enquiries back to the single point of contact. From there on, it is important that every other party be aware of the enquiries, the enquiry sources and the responses. This is the real meaning of 'media management': not managing the media, but managing the co-ordinated, considered and practical response to media enquiries so as to give a true and responsible, mature reflection of a given situation. This protects against adverse, usually unfounded and emotive reporting on a given accident or safety-related situation. Move beyond this knee-jerk scenario and the construction industry has every opportunity to develop and protect its positive safety record and situation. Nowhere can this be demonstrated more obviously than in real, everyday, construction scenarios where the

use of the media can have an instrumental influence upon an organisation's safety performance.

Local press and media

The media must be recognised as one of the most powerful tools in the
campaign to prevent accidents and injury in the construction industry.
Everyone has anecdotes about situations or events where the media has
damned an organisation's safety performance. However, occasionally
there is anecdotal evidence that involvement of the media in a real,
operational situation has contributed enormously to preventing accidents
and enhancing safety.

The assistance and co-operation of local newspapers can play a vital
role in the battle to avoid accidents, as well as in communicating safety
messages as widely as possible.

This can be exemplified by real situations where a contractor's plea to
the local newspaper helped prevent serious injury. A civil engineering
contractor involved in constructing major coast protection/sea defences
on the north-east coast was importing rock armour in the form of 7- to 10-
tonne rocks from Norway. The method employed was to have a large,
20 000 dwt barge anchored some two miles off the coast, with coasters
feeding this from Norway. To bring the rock ashore, a smaller 'feeder'
barge was used, floating as far up the beach as possible during the high
tide, discharging the rock into the shallows and then retreating to the
'mother' barge to reload. As soon as the tide receded, several backacters
(track-mounted excavators with large buckets or grabs attached to the
booms) were used to retrieve the rock and take it back up the beach for
placing at the foot, or toe, of the new sea-wall revetment. At that critical
time, the rock that had been dumped overboard was in very precarious
mounds in an extremely unstable and dangerous position. The danger
was compounded by the use of very large, fast-moving, heavy plant
moving large boulders up and down the beach from waterline to beach-
head.

Despite every kind of warning sign placed strategically around the site,
all along the promenade and at every beach access point, the contractor
had to contend with scores of children and youths clambering among the
rocks as soon as the tide receded. Even with the presence of the backacters
and their banksmen, the youths persisted in dodging vehicles and
jumping all over the unstable rock armour.

The agent concerned immediately despatched engineers, himself, the
resident engineer's staff and anyone else available to the local schools.
Headteachers were pleaded with to warn the children at school assem-
blies to keep away from the beach area. Parents and children were warned

it was at this stage of near desperation that it was decided to approach the local press. The public relations manager drew up the following press release which was approved by both the client and the consulting engineers:

'Nuttall seeks safety co-operation at Withernsea

The staff of Edmund Nuttall Limited, the main contractors constructing the new sea defences at Withernsea, are asking children, parents and schools to assist with safety on the construction site. Schools have already been contacted, but the plea for co-operation with Nuttall needs to go to all children and their parents in the area.

Since deliveries of large pieces of rock armour by barges to the beach commenced a few weeks ago, many children have been playing in and around the rock stockpiles. This has often happened while heavy rock-handling machinery has been operating close by.

Any construction site can be dangerous, but when piles of seven- to nine-tonne rock are being moved around the beach by large plant, the potential dangers increase. Children must *not*, in any circumstances, come near to such operations. Nor should they clamber around, on or between the rocks at any time.

Nuttall is eager to encourage all parents to keep their children away from the site and hopes that schools will also reinforce this message.'

The local newspaper was contacted by telephone and a journalist was fully appraised of the situation, both from a technical, engineering point of view and from a community safety angle. Then he was sent a copy of the press statement shown above. Finally, the newspaper was invited to send a photographer to the site to show, at first hand, the dangers involved on the beach works.

As a result of this campaign, the following newspaper article appeared on the front page of the *Houlderness Gazette* the next day:

' "Don't dice with danger" – Children warned

Curious schoolchildren who are crowding around the construction compound and playing on the rocks being used to defend the coastline are dicing with danger, and the company undertaking the works are asking children, parents and schools to assist with safety on the construction site, and are pleading for their co-operation.

Edmund Nuttall Limited, the main contractors constructing the sea defences at Withernsea, stress that any construction site can be dangerous. Their spokesman, Alan Smith, said: "Since deliveries of large pieces of armour rock by barges to the beach commenced a few weeks ago, many children have been playing in and around the rock stockpiles. This has often happened whilst heavy rock-handling machinery has been operating close by. When piles of seven to nine tonne rock are being moved around the beach by large plant, the potential dangers increase."

New signs warning the public of the dangers and effectively closing the beach were erected on Tuesday, and Nuttall says children must not, in any circumstances, come near to the heavy machinery, nor clamber around, on or between the rocks at any time.

Local schools have already been contacted, and Ian Hunter, Chief Engineer on the site, is planning to visit them soon, taking along a safety advisor to explain the dangers, "But the plea for co-operation", says Nuttall, "needs to go to all children and parents in the area now". Already 25,000 tonnes of rock are on the beach, which means just under half of the rock for this phase has been delivered. There are 8,000 tonnes in the exclusion zone and 5,500 tonnes on the waiting coaster. A further 20,000 tonnes will start being loaded on Thursday, which should arrive off Withernsea early next week, weather permitting.'

Luckily, this seemed to go a long way towards alleviating the situation. Certainly it did not stop the hard core of determined and totally stupid youths persisting in their irresponsible behaviour, but for the vast majority of them it seemed to work. In particular, the smaller children desisted. The local newspaper had brought the message home quite forcefully to the parents of those involved.

There are many other anecdotal situations where appeals to the local press can help a site team to keep the public away from a potentially hazardous site. One particular instance involved a public right of way and a road access from a small town and a vast adjoining caravan site close to a major resort in North Wales. Here, the contractor was constructing a major viaduct over a river and this access road as part of the new town bypass scheme.

In this instance, the public attending a public exhibition at the start of works on the scheme were fully warned of this potentially hazardous interaction of site operations and public pedestrian and vehicular activities. Posters and warning signage were erected everywhere practicable. Finally, leaflets were prepared for the caravan site management to give out to every holidaymaker arriving on site, warning them of the dangers of crossing the construction site and requesting no one to go outside the cordoned-off areas.

.... it became apparent that every single concrete pour, bridge beam placement and bulldozing operation had to have a huge audience, sunbathers in shorts and flip-flops avidly giving advice and, where possible, trying to help operations among the helmeted, steel-toecapped and reflective-vest-armoured construction workers.

Again, only by regularly and frequently contacting the local radio station and the town newspaper was the situation kept manageable during the critical summer months of vacation time. Press releases were issued warning of imminent operations, highlighting the dangers involved and exhorting all concerned to keep away. Regular interviews were given to the local radio station whenever a new or major operation was about to be undertaken. This, combined with greater co-operation with the police and security services, helped reduce incidents to a minimum.

Involve the press as early as possible

The local press should be involved as soon as possible in any construction project, and, at the first possible opportunity, the safety message needs to be pursued. The very first turf-cutting ceremony, where the local journalists and photographers have a field day with local dignitaries such as mayors and councillors posing for pictures next to the contractor's excavator, should make safety manifest. Signage should be prominent everywhere, especially on the safety posters that should be thrust into the hands of the VIPs as they smile for the cameras.

The contractor and consulting engineer, preferably with the client, should use this first opportunity to make contact with the local journalists and photographers. They must impress upon them the need to highlight safety issues, and must offer their full co-operation in the effort to prevent accidents. Only by having a greater awareness about the construction project that is under way will the journalists, and therefore the public, be better educated in safety situations.

If all this proactive work with the local press fails to work through the editorial medium, then, if a situation on a site becomes critical with public ingress and dangerous or hazardous incidents about to occur, the use of local advertisements needs to be considered. If the situation is serious enough, the professional and genuinely conscientious contractor would reach into his pocket and pay for a large, attention-grabbing advertisement to appear in local newspapers warning people to avoid a certain location, event or whatever the particular circumstances may be. The price of safety is incalculable, as is the cost of injury.

Technical press

Chapter 5 dealt in great detail with all aspects of the press: both local and national, as well as the technical press. Nevertheless, a short mention here will help to reinforce the importance of the technical press in the promulgation of safety issues to the construction industry.

It is most important that, whenever dealing with the technical press, due recognition be given to safety, whether this be a site visit by a journalist to undertake a major feature, or simply a news story or photograph picture caption. Wherever possible a company's utmost respect for, and full adherence to, all safety matters should be pushed to the forefront of all briefing material or interviews given to the technical press.

Unfortunately, certain sections of the technical construction press have appeared in recent years to have moved down the route of tabloid newspaper journalism when it comes to safety issues. For journals that are supposed to inform and promote their own industry, it is amazing how many times journalists will spot the slightest potentially or perceived danger in a situation and then blast it across the front page without either researching the situation further or seeking an explanation first from the company or companies concerned.

Often, the odd photograph shown out of context or cropped in a certain way can imply a terribly dangerous situation which, in reality, is perfectly safe. For example, as scaffolding is in the process of erection there may momentarily be no toeboard, a half-fixed hand-rail or a ladder propped temporarily to one side without being tied. A photographer could take any number of pictures on any construction site showing dangerous or potentially dangerous situations or potentially hazardous materials, plant or tools. But the vast majority would have a rational explanation, namely that they were in the process of assembly, removal or other ongoing activity.

These photographs, taken out of context and then processed back at the editorial offices, can end up on the front page of the construction journal concerned, with a caption written by a news editor or sub-editor who has never been to that site or seen the context in which the picture was taken. He or she has usually had no contact with the photographer concerned either; so again there is no cross-referral to the site itself. These scenarios can combine to show the most scandalously dangerous situations that actually have no bearing whatsoever on reality.

Yet those same reports can do immeasurable damage to the organisation(s) concerned. The whole industry will raise an eyebrow if it sees safety regulations being flouted . . . or perceived to have been flouted. The Health and Safety Executive can often become involved and instantly undertake a full-blown HSE site inspection. Clients would be alerted to

One public relations manager was telephoned by a senior, respected and respectable news editor of a leading civil engineering/construction journal who has since left the magazine concerned. The manager was informed that the journalist had in front of him a photograph showing a highly dangerous situation. The public relations manager, sitting in a car in Leeds and about to go into an important meeting, asked for a description of the picture and where the site was. The response was, simply put here, that a labourer was standing on a wall wielding a sledgehammer in the process of demolishing that same wall, with neither harness nor rails, nor any other form of safety protection. The site was in South London.

Knowing the site well, the public relations manager explained that it was quite likely that this was another contractor since several were involved in and around and adjacent to the same site working on other albeit related aspects of the same, large project. However, he asked for time to receive a copy of the picture to identify the contractor concerned and to see if there was an explanation. The journalist concerned refused and said he would be publishing it in the next issue and was simply asking for a comment or a reaction to this allegedly dangerous situation.

In these circumstances, the journalist not only showed irresponsibility but also gave the company spokesman no chance to defend the situation. The spokesman could only say that in the circumstances it would be impossible to comment upon a picture he could not see, nor would he comment when he did not even know if it was his own company involved. He concluded, quite rightly, by saying that the journalist could publish the photograph if he liked (because he could not be stopped) but that it was totally unfair to expect any kind of constructive defence whilst he was sitting in a car 280 miles away.

Luckily, the journalist, after much cajoling, agreed not to use the photograph but said he would keep it in the photographic files specifically concerning site safety and would, in due time, use it to show bad practice. When, some time later, the public relations manager received a copy of the photograph, it was accompanied by a note from the journalist reiterating that this would be used at some future date and the site and contractor would have to be mentioned. On sight of this photograph it was obvious that another company was indeed concerned, and it was a subcontractor's labourer working on a small, totally unconnected project alongside the main scheme.

Think of the damage that could have been done, *wrongly*, to the company falsely accused of responsibility for the dangerous situation in the first place. Its reputation could have been harmed irreversibly by such publicity. No amount of letters to the editor or any one item in the Erratum section would have eradicated the damage concerned.

There are numerous other anecdotes that could be used to show how very harmful the technical press can be to a company's reputation in the context of safety. Moreover, there are many, many examples of journals using photographs in their pages which lead misguided, misinformed observers to rush out a letter to the editor pointing out a hazardous situation that they have picked up, again out of context, in the photograph concerned. In nearly every case there is a logical, proper reason why that situation *appeared* to show some compromise to safety but in reality, posed no threat whatsoever to anybody's safety.

All of this merely reinforces the need to be extra vigilant when the technical press come anywhere near a construction site. If a request to visit a site is received from the construction press, the simple advice is to say no. That is, no in the immediate future, but that such a visit can be arranged given proper, manageable time. There are many reasons why impromptu, immediate visits to sites where the personnel concerned have not been given sufficient forewarning, should not be entertained. The main one is safety, the safety of the journalist and/or photographer concerned as much as anything else.

Before a journalist visits a construction site it is advisable that a safety officer undertake a very thorough inspection beforehand – preferably two visits: one to identify any aspects that need addressing in general terms (such as site tidiness, since an untidy site is always potentially a dangerous site) and a second visit immediately prior to the arrival of the journalist. This latter visit can pick up those situations referred to above where ongoing, in-process works lead to a temporary *perceived* potentially dangerous situation, for example the case of the temporarily propped ladder not in use but left to one side and, from a safety regulation point of view, dangerous if not tied into place.

The engineers or site managers that will be involved in any interview with the technical press also need to be briefed beforehand to ensure that they give due mention to safety procedures in hand on the site and the importance of safety in any and every situation within the construction programme. They also need to point out all safety features as they take a journalist around a site.

It is equally important that any background briefing documentation that is pre-prepared for the journalist or journalists concerned gives due recognition to safety issues and safety systems or regimes in place. For example, if the project concerned involves strengthening a bridge or viaduct where the steel box sections need additional steel reinforcement to be welded on to the inside faces of the box girders, special mention

pairs, not as individuals; fresh air pumping and venting on a continuous basis; air monitors in place; and safety induction talks prior to work starting.

Thereafter, the matter is entirely in the hands of the journalist. If a site really is 'squeaky clean' from a safety point of view, and it should be anyway, then there should be nothing to worry about from a technical press visit to the site.

The technical press can also, like the local press (covered above), be very helpful in promoting safety issues. If a company has developed or evolved any aspect of safety that has led to a safer environment, then it should be publicised – not only because it shows a company in good light for evolving such a method, but also because it can let others benefit and so reduce accidents within the industry at large. Responsible journalists from the technical press should always be interested in covering any safety issues from which lessons can be learned by others.

Whether it be a new safety product, service or work practice, any advancement in avoiding accidents should be publicised as widely as possible. Once it has been explained to the technical press, then it should be given prominent coverage in the appropriate sections of the paper/journal.

At the end of the day, the technical construction press should be promoting the industry. The more support the public relations department can give to the industry in enhancing safety issues in the technical press, the more quickly and effectively will the industry be able to address safety more seriously and more tangibly. Think positively and be proactive in pushing safety issues when dealing with the technical press. Only good can come from such an approach – for the company concerned and for the industry at large.

Communicating safety with the local community

Generally, the media's greatest input to safety relates to the public at large, particularly the local community adjacent to or close by a construction site. By definition, therefore, it is the local press, television and radio that are most important in pursuing safety issues at site level. And the construction industry need not be so defensive about dealing with local press; it can work with the media to promulgate safety surrounding site operations.

There are numerous situations where the safety message can be put over to people in the surrounding area. These can usually be initiated from award of contract and prior to start of operations and then continue throughout a project's duration until the final completion ceremonies. Each stage – turf cutting, construction milestones, completion/opening events – can give ideal opportunities for a company to reinforce its safety messages.

As soon as a contract has been awarded, that should signal the need to identify any possible situation to begin the safety communications strategies for that project or scheme. Whether the site is in an urban environment or a green-field, rural area, there will always be a local population that will be affected by the works. A first priority should be to liaise with the client's public relations practitioner to identify what issues are prominent locally and where potentially hazardous interaction between the public and the construction works could overlap.

The client always knows best when it comes to the local community. Usually, the contractor and his staff are newcomers to an area even though they may be 'local' in the broad, regional scenario. Not only does it reassure a client if a newly appointed contractor immediately makes contact locally about safety issues, but it also gives the project manager a very early opportunity to identify potential safety situations and enables him or her to prepare accordingly.

The public relations function in this context can be critical. It can give the project manager or agent an immediate list of elements to be addressed from day one of a project. For example, how many schools are there in the vicinity of the works? Where are public footpaths or rights of way across or through a site? Where are the most prominent vehicular/pedestrian interaction zones on the site? Upon which members of the community will noise, mud, dust and other potentially dangerous media be imposed during the construction works?

Most of this will be contained in the site safety plan. However, a proportion of the 'indirect' impact of the works will not. Only the public relations element of the works will pick up these potential aspects where *safety* of the public can often be confused with *appeasing* the public or patronising it with generalities about noise, vibration, increased vehicular activities and so on. Each of these 'indirect' impacts can be possible safety issues.

The turf-cutting ceremony

The turf-cutting ceremony for any scheme can be an important marker for the ongoing safety strategy. The local mayor, member of parliament, secretary of state or minister will often be involved in such a ceremony. They are there because of their prominence and because they represent the local community. Their presence should be capitalised upon. Use

project, not only for the on-site works and workers, but also for the local community.

Do not be afraid to ask the local politician at the photo opportunity to hold a site safety signboard reading 'Construction sites can be dangerous. Children keep off!' They are usually more than happy to do so because it does their own public relations an immense amount of good as it reinforces their own commitment (or perceived commitment) to the safety of the population.

Turf-cutting ceremonies should also be visibly and loudly proclaiming safety throughout the organisation of the event. Insist that all participants have easy access to the ceremony site, clearly marked with hazard warning tape or bunting keeping them from the general construction site activities. At the earliest possible opportunity (preferably even before entering the site compound), insist that everyone – including the loudly protesting press photographer – wears a helmet and high visibility reflective vest, regardless of sex or hairstyle! Segregate the party at all times from any vehicle that may be performing a turf-cutting action; usually a backacter or excavator. And cover the place in safety posters and signage. All will show the press and media that the contractor is professional and means to enforce safety regulations to the highest standards.

It is also at this stage that the contractor should commit themselves to safety on the site to a ready-made wide audience as well as to the media attending. Traditionally, the politician, whether local councillor, mayor, MP or minister, will be introduced and will undertake the first speech. The client or developer will follow with details of the scheme and then formally hand over the site to the contractor for works to commence; whereupon the contractor invites the politician back to mark the start of works by digging/excavating the first sod or turf. That small gap for the contractor is an ideal opportunity to push home the safety issues.

The contractor's representative at such an event is usually selected to match the importance or size of the contract and client as well as to reflect or complement the seniority of the politician or turf-cutting VIP. The more senior the contractor's representative on the day, the better. He or she can speak with more authority and lend increased credence to the safety commitments given. Wherever possible, the public relations practitioner should either insist on his company's representative including safety in his or her speech, or should write it into the speech if that role is performed.

However short the speech, the contractor should refer to the potential dangers that any construction site represents and should commit the company to being as safety conscious as possible. The speech should go on to ask for the full co-operation of the local community, particularly children. It must urge them to respect safety signage and to keep off the site, even when it is quiet at the end of the working day. Finally the speaker should request the help of all present in promoting that message.

In any subsequent interviews with press or other media at the event, these same elements should also be reinforced to the journalists or interviewers concerned. Any mention of safety aspects will help to create an image of those undertaking the works as responsible and safety conscious. That in turn should reassure the local population that specific attention is being paid to their safety as well. Moreover, if the local politicians take these messages back to their colleagues, they will be promulgated and reinforced even more widely.

Once the excitement of the start-of-works ceremonies is over, the serious application of all those promises of safety has to be implemented. The spotlight will be on the site and on the ability and willingness of the site staff to deliver. Thus, from a public relations point of view, every effort needs to be put into the immediate erection of applicable safety signage throughout the site and its boundaries.

Open evenings and other public forums

Coinciding with the start of works or, even better, before construction works actually begin, the local community should, where possible, be given the opportunity to understand the scheme in operational terms and to learn about the key phases of the construction programme that will affect residents: when noise/vibration from piling will start/end, when diversions will be put into place, when and where pedestrian diversions will be required, and so on. The organising of a public open evening can be one of the most effective ways of communicating such information. At the same time safety issues can be highlighted and brought to the fore.

Through co-operation between the client, consulting engineers and contractor, the organisation of such a forum to inform the public can pay enormous dividends during the construction period. An advertisement in the local press, as well as notices put up in the local library (and, where practicable, local newsagents' windows and the like) should give warning and encouragement to attend such an event. The notice should specify that the nearby construction project is about to start or has just started and that engineers will be available to discuss issues surrounding the works.

It is therefore essential that senior and well informed representatives of all three organisations be available for comment at these occasions. All questions raised by the community should be predicted. If an answer cannot be given immediately, a mechanism needs to be established whereby answers can be found very quickly. Such an event is not a public enquiry, nor should it be used to vent local opposition or apathy to a

tions for the public and these need highlighting. Where better to bring such anxieties or concerns to light than in the local village hall or library where on-site engineers can promulgate their concerns about certain elements within the scope of their works? The client will also be aware of local safety issues and can help feed genuine concerns back to the contractor and consulting engineers so that specific areas can be addressed.

If, at such an event, presentations are given by the site staff to the audience of local representatives, then the profile of safety issues needs raising. Likewise, if scheme plans are part of any exhibition displaying the scope of the project to the public, then any potential accident blackspots such as site vehicle/public interchange or public footpath/site overlap situations should be highlighted and the public warned strongly about such concerns.

Public open evenings or exhibitions also allow the local environmentalists or community trouble-shooters to benefit from an open and frank discussion about issues surrounding a construction site. Should there be any such issues, there is also the opportunity for site staff to bring their influence to bear upon the safety ramifications, particularly as regards site security and the need to keep people out of operational areas for safety reasons.

Although it is preferable to organise such events as early as possible in the programme of works, it is usually difficult to do so within the very limited time available between contract award and full mobilisation. However, part of the mobilisation schedule should include provision for a public information programme of some description. It is equally important that, once such an event has been held, an opportunity should exist for further consultation – particularly if a construction project has experienced safety-related incidents or potentially safety-compromising situations.

It could be that the project manager or agent has the opportunity to address meetings of the local Chamber of Commerce, the Mothers' Union, the Round Table or the Working Men's Club Forum. This should be encouraged because not only will it familiarise local people with the project in hand and its progress, but it will also give ample opportunity to raise safety issues among local observers and the public at large.

If a project is of a long-term nature, then public open evenings should be encouraged at interim stages to update people with progress and prominent issues. Certainly these events should be encouraged if there have been significant safety issues arising from the works. However, if the project is of a short duration or is uneventful in terms of safety issues, then the final 'topping out' or 'tape cutting' ceremony should be marked by comments about the successful safety record of those involved in the completion of the works.

Again, the speech writers for the VIPs performing the completion of works ceremonies must, wherever possible, publicise the safety achievements of a particular project where such praise is earned. They should also use the opportunity to wish a safe and successful future life to the end product, whether it be a road bypass, a new superstore or an office block. The end user, be it driver, shopper or office worker, needs to be assured of safety at all times – especially by those who constructed the facility in the first place!

Communicating with schools and children

Throughout the process of any construction works, whether they be connected with a supermarket complex, a new housing development, a major road bypass scheme or an office development, a major aspect of community relations must be concern for children and school liaison. Adults should be able to comprehend basic safety issues and situations. Children cannot necessarily be expected to understand.

Nowhere can the overlap between the safety department and public relations department be so highly prominent or productive than in schools liaison and children's safety. Getting the safety message over to children has very wide repercussions, not only in promulgating safety issues throughout a school to children, school teachers and staff but, equally important, to siblings and parents or other relations at home. Real, genuine and committed programmes for school safety liaison need to be established by any construction company as soon as it wins a contract within the vicinity of a school or a young community.

School safety talks

At the time of the first turf-cutting or start-of-works ceremony, efforts should be made by the contractor to communicate with the headteachers of every school in the proximity of the ensuing works. Generally, the site staff should contact every school that will be affected, directly or indirectly, by a construction scheme, and offer safety talks to every age group by qualified safety officers connected to the site. This needs to be sold very strongly to school staff and actively encouraged as a genuine attempt to educate children in safety issues, illustrated by real-life situations adjacent to their own homes or school.

Not only will school safety talks help the contractor on site, but they will also paint a picture of a much more professional and responsible civil

children key messages and the ability to reap ongoing educational benefits from those initial presentations or talks.

First and foremost, safety officers are teachers: they must educate, advise, and reprimand for failures. In this respect, they are usually very good at managing their own safety presentations at schools. But public relations practitioners should be equally adept at giving over the safety message to children, reinforced by the safety officers. Between them there is a recipe for an extremely effective presentation to children that calls for both presentational flair and pragmatic message giving.

There is no predetermined way to undertake these school safety visits: each one will vary according to the ages of the children, the nature of the construction works being undertaken and the character of the presenter(s). Nevertheless, since the concentration span of small children is so short, it is wise to be thoroughly prepared in advance, with a multitude of fast-moving activities or subjects to keep up a constant interest.

The use of two or more presenters also helps keep the presentations moving at a suitable pace. In the examples given earlier in this chapter, the public relations and safety officers interacted throughout the presentations, with the public relations representative undertaking the introductions, the presenting of safety signage and posters as well as putting down questions and answers to sow the seeds of basic safety awareness before handing over to the safety officer. The latter then described his function within the company, showed how construction workers protect themselves (including dressing up two volunteers in helmets, high visibility vests, steel-toecapped boots, gloves, goggles and ear protectors!) and then identified specific areas of potential danger or hazardous situations relating directly to that particular site or locality.

Involve local police

It can often be useful to involve the local police schools liaison officer at such presentations. Although there is a danger of stretching the presentation over too long a period (half an hour should be the maximum amount of time, including questions and answers), the addition of a familiar face or an obviously influential 'establishment' figure can help reinforce and formalise the safety messages being given over.

The visits to school also have to be educational, with an ongoing life. The use of safety posters and quiz sheets for the teachers to keep helps generate continued safety awareness over the days and weeks to come. Likewise, the donation of helmets and high-visibility vests allows teachers to continue the theme indefinitely in role-play situations in the class. The issuing of pens, pencils and notepads suitably adorned with safety messages also helps

One of the main benefits of such activities is that the children then take the safety messages home to parents and siblings. This in turn helps promulgate safety further into the community surrounding a construction site.

The project manager or agent must then be strongly encouraged to follow up any goodwill fostered with schools. It never does any harm to offer the construction site and operations as a suitable school project. Painting competitions can be organised with prizes being given by the project manager. Site visits can occasionally be organised so long as the maximum amount of protection is given to the children. Thus they may need to be confined to a minibus while going around a site.

The value of such interaction also contributes enormously to the contractor/client relationship. If the client can see the full and active involvement of the chosen contractor with the community, showing genuine concern for the safety of children and the public at large, the benefits will extend into the project itself and reflect positively on the client. If, on the other hand, there is an obvious disregard for the need to promulgate safety issues within the community, then that attitude will soon reflect upon the integrity and image of the client. Good experiences in this respect can help no end in the follow-up orders or ongoing pre-qualification with a client. Both client and contractor can also use positive community relations and safety issues in their project presentations or award applications following project completion. Thus they are all part of good business practice and help business success.

At the end of the day, the whole objective of this aspect of safety in construction is to avoid or limit the ingress of children on to construction sites and stop accidents and injuries. None of us in the industry wants to see articles like this, which appeared in *The Times* on 21 March 1997:

'A 14-year-old boy suffered leg injuries when he was run over by a mobile crane after he and five other teenagers broke into a Gloucester building site and drove the machine around. He had severe cuts and was taken to Frenchay Hospital, Bristol, where he was in a satisfactory condition.'

Summary points

- Safety makes a clear contribution to the reputation and image of construction organisations.

the organisation. Public relations' role is to communicate, promote and reinforce safety issues.

- A key objective is to promote safety messages to the public surrounding, or affected by, a construction project.
- All objectives are linked inextricably to the legal obligations of the industry.
- Communications in the form of safety newsletters, press and media campaigns, school and community liaison, posters and external marketing communications must convince clients or potential clients of a company's commitment.
- Public relations must be involved in the induction of new recruits, in training programmes, and in the showing of HSE videos and the organisation of safety competitions.
- It is vital that a main contractor has the full cooperation of the sub-contractor, and this requires communication.

Chapter 12
Crisis Management 'That's another fine mess...!'

Introduction

Any crisis or incident in the construction industry is potentially disastrous, not only to people but also to the company and the industry itself. The previous chapter on communicating safety in construction demonstrates that the construction industry, no matter how carefully and well established the safety and health regulations and rules are, represents a very dangerous business. An incident can quite easily and quickly become a crisis.

The setting up of a crisis management system or mechanism is one of the highest priorities for every company. The system must operate not only as a generic formula, but also on an individual contract-specific basis for those projects identified early on as potential incident-sensitive sites. Moreover, a crisis plan is not a public relations scheme. It involves top management; the personnel, safety and public relations departments; the project manager; and interaction with emergency services (where applicable), clients, consulting engineers/designers, local authorities, the press and media, and so on. Indeed, crisis management is essentially an exercise in the effective, controlled flow of information between many different parties.

There are numerous books and other sources of information available on crisis management. All of them have a very valid and useful role to play in the planning and establishment of a crisis management scheme. Here we intend to look more specifically at the evolution and then control of incidents or crises in the construction industry itself.

What is a crisis?

We must define what a crisis is in the context of the construction industry, and say how this is different from an incident. Every day, on virtually every construction site, incidents occur and are managed without any

damage or hazard, or relates to the company's finances, reputation or otherwise. But 99.9% of all such, usually minor, incidents are immediately controlled, overcome, managed or alleviated. A very tiny minority of site incidents can become crises which affect a much wider audience, involve many more participants and outside parties (direct or indirect), and, usually, result in media interest.

An *incident* could be a near miss such as an operative just avoiding the slewing tracks of a D9 bulldozer or the fall of a diesel drum that remains intact. A *crisis* could develop if that operative was actually caught and injured or killed by that D9 truck or if that drum of diesel had burst open and emptied into the adjacent river. In the first case, the near misses must be reported to avoid repetition and to enable a mechanism to be set up so as to prevent repetition. This is incident management. In the latter, crisis management takes over as the information about the incident begins to spread to a broader audience.

How can we manage crises better?

In this section we examine briefly some of the more practical elements of establishing both company-wide and site- or contract-based mechanisms for crisis management. Before looking specifically at how in-house resources (both safety department and public relations department as well as site staff) can be organised to cope with crisis information, it is worth looking at some of the outside resources available.

PR Week (1997a) featured an article entitled 'Protect yourself from a PR crisis'. This discussed the launch of a new insurance package in which an insurance company was offering to provide professional crisis support as part of its policy cover for a large business. The article stressed the need for all businesses to have insurance policies against such events as fire or theft. Public relations crises represent a less obvious financial risk, but can be equally damaging to an organisation. 'So the comfort of having crisis communications support included as a benefit on your insurance policy is undeniably appealing' (*PR Week*, 1997a).

The article went on to make the very valid point that a major danger of such insurance policies is that it may make the policy holder assume that this is the only step he needs to take in order to protect himself against a crisis. It is not. While having a team of public relations fire-fighters on hand to help an organisation through the first few days of a crisis may be comforting, it is far better to avoid the need to call them out in the first place. In other words, prevention is far better than cure.

In this sense, put into the context of the construction industry and the variety of incidents that might evolve into crises, the conclusion given by PR Week (1997a) is equally applicable to our business.

'Every organisation – without exception – faces a clutch of issues and contingencies which could, in certain circumstances, develop into a full blown, bouncing off the walls crisis. Managing and monitoring those issues can often prevent them from doing so...
... At the very least, they should have a tried and tested plan for crisis PR handling in place to tide them over until the cavalry arrives. More importantly – and perhaps even as a condition of their insurance – they should take some preventive public relations medicine of their own.'

This illustrates two main points. The first is that insurance policies, however cognisant of emergency or crisis situations, cannot prevent crises, nor do they really provide cover for an organisation's reputation in the world, its ethos and standing amongst peers, which can be affected by a crisis scenario. Secondly, having a hired-in PR agency to handle such a situation may help in fire-fighting during a particular event, but again does nothing to assess issues and events beforehand to identify potential trigger points to a crisis and thereby contribute to prevention.

Most potential crises in the construction industry relate to safety or health aspects, either directly or indirectly. Picture the scene of a collapsed tower crane; the failure in high winds of a whole wall of scaffolding collapsing into the streets below; a crawler crane overturning and catching overhead cable; a major tunnel collapse. All failures affect lives in these situations. And how do we cope?

Unfortunately in this age of reduced funding for fire services, the health service and local authorities, any incident can be used by these bodies to maximise upon funding deficiencies by highlighting how, against all odds, they have saved the situation. 'Unfortunately' is the appropriate word, because the telling of the story of the events often ends up in melodrama created to maximise publicity but which becomes far more removed by the minute from reality. Quite frequently we see the local newspaper quote the local fire brigade officer as saying that 'The rescued party fell 40 metres from scaffolding into a pit, and [thanks to this and that] survived and is recovering from multiple injuries in hospital'. Upon further investigation we find that someone has inadvertently stepped off some ladders and tripped 2 metres, twisting his ankle and banging his head (helmeted of course) on the ground – and is back at work again before the paper is even published the next day! An incident, in local newspaper terms, is a crisis in the eyes of the Health and Safety Executive and the company involved.

ready to react strongly to greatly exaggerated reporting that sensation-alises minor incidents.

Thankfully, most construction companies actively pursue the most rigorous policies of accident prevention. Every responsible company aims to either eliminate or reduce to a minimum any risks arising from its work activities. It must be acknowledged, however, that, despite those measures, major incidents may still happen. Again, generally arrangements exist for the more likely emergency situations, i.e. injury, fire, work over water. These may be modified to suit particular site conditions on the basis of further risk assessment.

Where work takes place on sites for which emergency procedures are already in place, such as chemical plants identified as Control of Industrial Major Hazards (CIMAH) sites, or sites for which emergency procedures are in place and dictated by the client, typified by the railways industry, the company should ensure that these procedures are in place and are fully implemented.

Such procedures, allied to responsible safety training, are intended to prevent or minimise any accident occurring on a construction site. As shown in Chapter 5 on media relations, in the event of an incident the observer's first question is inevitably: 'Has the company done everything it could to prevent the tragedy?' If procedures are in place, then this question can instantly be answered in the affirmative.

Incident management planning

Before producing any incident management plan, therefore, the above question needs to be answered. Only when an assurance can be given that full procedures and training have been implemented can a plausible communication mechanism be set into motion.

So what is an on-the-ground, practical incident management plan? In the context of a construction site and the wider company that is responsible for that site, much of the theoretical, academic approach to incident management can be discarded as hyperbole or as relevant to other industries or scenarios. At the end of the day, a site and its staff need a very basic but effective *communication plan* that can fall into place alongside their own safety and emergency procedures together with details of the local accident and emergency unit, the local police station, first aiders, line management/safety officer reporting and the like.

Thus a site manager or agent has to have pre-briefing, verbal or written, about incident management plans. It needs to be explained that incidents

are events such as accidents that attract public and/or media attention
outside the normal contractual pattern and often in an unwelcome way.
By their nature, incidents are unplanned but their management has to be
planned, both practically and in public relations terms.

On this basis, most construction companies have a generic incident
plan that is issued to every site agent or contract/project manager.
Detailed planning is contained in the relevant site health and safety plan.
The principles according to which this is given effect are formulated in
advance by observing the following measures:

- Agreeing and naming one person to represent the client who has
  principal responsibility for statements, responses and news releases to
  all relevant media.
- All statements involving the company name, directly or indirectly, are
  agreed beforehand with the public relations manager who will obtain
  approval from all relevant line managers/directors.
- There is a prearranged sheet of telephone numbers, dedicated telephone lines, etc. to facilitate and control co-ordinated external communications.
- Pre-acceptance of the following holding statement as the maximum
  authorised to be given by agents or safety officers who may be
  required to respond to the media immediately and before other
  arrangements are in place.

STATEMENT BY . . . . [company name]

It can be confirmed that an incident has occurred at this site. However, until a
full investigation has been undertaken by the relevant specialists, we are
unable to make any comments about the situation. To comment at this stage
would be entirely speculative so we cannot answer any further questions until
the relevant authorities have had an opportunity to fully investigate the incident.

   If, at a later stage, you wish to obtain any further information, please refer
your questions to our spokesperson . . . . [name]

The statement is the minimum, immediate reaction 'equipment'
required by site staff. It is a standard, one-sheet guideline that is easily
referred to and is unequivocal in its action requirements. However, the
most important element is the appointment of a representative as
described above, particularly for a site where, in advance, any out-of-the-
ordinary events or issues can be predicted. This could be a contaminated
land rehabilitation site being remediated close to a residential area. Or it
could be a controversial bypass construction facing opposition from a

Whatever the site, if it has been identified as a potential incident-susceptible contract, or if it has a particularly sensitive client, there should be an *incident communication plan.* On some sites this may be referred to as an *incident flow chart* or a *third-party questions flow chart* or similar. In each case a single A4 sheet is the contact sheet for communicating with every party involved in a contract, with individual names, telephone numbers, e-mail addresses and titles, organisation names, etc. In essence it takes any site-based incident and initiates a flow of information. It provides a cross-referral and approval mechanism whereby all parties are kept fully informed by a spokesperson for the whole network who is released to respond to outside enquiries.

Usually, the flow chart begins with the contractor on site, with the agent or project manager issuing information in two directions: internally to the public relations manager (who is part of a network that links him or her to the managing director and other parties); and to the client/consulting engineer side. This initiates a parallel flow between the public relations manager of the contractor and that of the client or its public relations agency should one be employed. Within this flow, the appropriate response is agreed and the previously agreed spokesperson is unleashed.

In some contracts the flow chart will accommodate two or more separate but interconnected communication routes for different types of incident. For example, the flow chart may begin not with an incident *per se* but with a query: from a member of the public, from an adjacent property, or from the press – each having a slightly different and more appropriate routeing.

In some very high profile projects that are the focus of constant attention from the national and local press, radio and television, a fairly complex incident flow chart can arise. In particular, a large development may go well beyond the site and direct participants at the time, and reverberate to participants far and wide. These could include local or national politicians, financial institutions, the developers themselves, the potential retail end-users, lobby groups, sub-contractors, major suppliers and, in turn, all their own prime contacts and public relations officers or agencies. However, it is important to keep the flow chart as succinct as possible, on one sheet of A4 with clearly defined lines of communication, with the priorities of those lines, and with specific responsibilities in the lines of communication.

The success of any incident plan depends upon the level of staff communications within a company (see Chapter 9: Internal communications) and upon how staff are guided to respond to incidents. The company manual issued to all site managers up to director level should contain guidance on all of the above elements. A valuable rule or company regulation that should be included in any staff induction documentation and any subsequent guidelines issued to staff is as follows.

'All matters with potential to adversely affect the Company's reputation or image are reported to the Public Relations Manager who, in turn, liaises closely with the directors.'

Undoubtedly there are very many elements that can be built into an incident management plan. However, if the preceding elements are built in at ground level, many of the initial stirrings of an incident can be nipped in the bud *before* they become a crisis.

The final element, which is vital to all of the background planning and preparation previously undertaken, is the attitude and behaviour of the most senior management, the directors, managing director, chief executive and/or chairman. Minor, irrelevant incidents may become 'bouncing-off-the-walls crises' through the behaviour and actions of the top representatives of the company. In the case of a minor incident, the presence of a senior director on the scene can create something out of nothing. On the other hand, a major incident involving either a fatality or a serious impact upon the community alongside a site should be given the respect shown by the presence of senior representative(s). If a company is claiming to be concerned, then that concern should be openly demonstrated by a senior presence.

These latter considerations must be built into any company incident management plan *before* an incident happens. The public relations manager needs to know from the start the position his managing director will take in an incident; and to be reassured that off-the-cuff, uncoordinated remarks will not emanate from that direction before all elements have been considered. Any public relations manager should therefore be able to go into the managing director's office to ask how he or she would deal with an incident or crisis in various scenarios.

On-the-day reaction

Incidents can turn into crises very rapidly. They are unplanned and usually totally unpredicted and may be responded to 'from the hip' in the opening minutes. An immediate *stop* needs to be put on the chaos and resort made to the incident plan, calmly and logically. Most of all, a thoughtful approach needs to be taken. Then, upon agreement of all parties within the incident plan, as much authorised information as possible needs to be given without prejudice to ongoing investigations.

There are few statements more annoying than: 'The company refused to

...ng understanding that premature conjecture can lead nowhere. It is not 'No comment': it means simply 'The company acknowledges a situation/incident has occurred and is doing everything possible to investigate that situation. It would be irresponsible to say too much in the meantime...'. This is a far better, more human response. Thus, the set statement is the best instant response available. Thereafter, the scene will unfold in a controlled, co-ordinated manner, if the plan is adhered to.

Not very long ago a major civil engineering exercise went wrong, and ended up on the national news on radio and television and on the front page of the national and local press. At the conclusion of the event, questions were asked about what was right and what was wrong about the press and public relations handling of the crisis.

What was right?

- Full pre-briefing of press and radio
- Good press facilities throughout
- Organised interviews with technical staff
- Escorted photo opportunities throughout
- Good intermediate statement advising that delays would result.

What was wrong?

- No advanced planning by those parties involved for the eventuality of a crisis
- No adequate, 24-hour PR presence on site
- Too many 'spokespersons', each directing press to others
- Not enough prepared, agreed statements
- Too many 'No comments'
- Lack of organised technical interviews.

Technically, and in civil engineering terms, the event was highly successful and revealed tremendous engineering skills and achievements. In public relations terms, both specifically *public* and *political*, it became a fiasco in the public eye and an embarrassment politically.

The main revelation was that the public relations managers involved had not at any stage considered in advance what could go wrong, how they would collectively or individually respond, or who would take overall responsibility for co-ordinating responses. Only a pre-prepared plan would have alleviated the situation they faced over a very tense four days, 24 hours a day.

The incident also showed the need to communicate, albeit only with regular 'statements' rather than hourly interviews. The lack of communication for long periods was intensely damaging. Stage-by-stage statements on what would happen next should have been agreed by the parties concerned and issued regularly. This is demonstrated in a non-construction scenario which recently affected the rail industry. The Southall rail crash could, for all intents and purposes, be a construction incident or crisis. In this case, as we would all expect, Railtrack had a very well rehearsed and organised crisis management system. It is a very good illustration for any industry, but especially for us in construction.

Following privatisation, Railtrack had a much more complex management structure, with Railtrack owning the tracks and separate operating companies running train services. Against this background, Railtrack had carried out several 'table top' theoretical exercises and on-site accident simulations. However, Southall was the first rail crash involving multiple fatalities and injuries. It occurred in September 1997.

Most observers agreed that Railtrack handled the crisis extremely professionally. As at a construction site where a range of clients, consultants, suppliers, subcontractors and so on are involved, Railtrack had to ensure that all of the different companies involved at Southall worked together in responding to the accident. Together with the wide variety of emergency services on the scene and off it, the company presented a coherent, caring and united front in giving out information against 'an emotional background of rampant speculation' (PR Week, 1997b).

The set, planned and rehearsed system proved to be very effective. The practice exercises had established a crisis management plan and the necessary lines of communication between all the companies concerned and the emergency services. The Railtrack Press Office itself became one of the central points of contact for industry comment and concentrated on feeding all the latest information directly to journalists as opposed to issuing press releases. But this was co-ordinated and information was vetted before release. Moreover, working in shifts, press officers fielded more than 5000 calls over four days.

The verdict of PR Week (1997b) was very positive:

'The press office responded swiftly to speculation ... with spokespersons on site and phone updates conveying a co-ordinated message that comment on causes of the crash had to await the enquiry.'

Other comments were that Railtrack had people available around the clock and showed itself to be a cohesive force. This speaks volumes for the need for a co-ordinated, planned incident plan that, if adhered to, can work wonders.

given here is of a practical nature intended to guide in-house public relations practitioners within the construction industry. Nevertheless, it would be advisable for every public relations manager to establish a relationship or rapport with at least one agency with a track record in crisis management. Should the crisis happen and you are by yourself, besieged by radio and television reporters, with an irate managing director who refuses to talk to either, you may need to call your pal from the agency and say, quickly and calmly of course, 'Help'. For a price, he or she undoubtedly will.

Summary points

- This is deliberately a short, sharp chapter. It intends to guide the reader into a frame of mind where it becomes essential to produce a real and practical incident management plan. This plan in itself will help contain the incident and stop it becoming a crisis.
- Crisis management is simply about communication in a comprehensive, co-ordinated and coherent fashion leading to the controlled dissemination of information that does not prejudice any resultant investigation. It is about control as much as anything, and therefore it is about having established guidelines and mechanisms that will lead to that control. All parties need to be informed in advance about such procedures and to know what role and responsibilities each has when an incident occurs.
- Without a set plan, agreed beforehand by all involved, a minor incident can very quickly snowball into a crisis. A short, sharp, simple plan, distributed to everyone, will go a long way towards controlling an incident.

Part Five

Despite positive attempts by organisations such as the Construction Industry Council (CIC) and the Construction Industry Board (CIB), the perceptions that people have of the construction industry are affected by the negative impact of the industry's processes and products. The problem is not just an issue of image, but also one of substance.

In response to the work of Sir Michael Latham and his report in 1994 on 'constructing the team' (Latham, 1994), the Construction Industry Board Working Group 7 recommended that the industry should develop a series of initiatives as part of an industry communications strategy (CIB, 1996). The aim would be to combat the 'cowboy' image of construction and integrate new approaches to marketing, productivity and cost improvement, careers, training and equal opportunities. In its report entitled 'Constructing a Better Image', the Working Group announced that the campaign would have a number of internal and external objectives. The internal objectives would be to provide better value for clients; to improve quality, professionalism, efficiency and profitability; and to enhance the professional relationships between constructors, consultants and clients. The external objectives, aimed outside the industry, would attempt to attract greater investment, improve environmental and social relationships, attract high-standard recruits and encourage equal opportunities.

As part of the strategy to achieve these objectives came the launch of the Considerate Constructors Scheme. An editorial campaign aimed at national and regional television and radio and the technical, trade and consumer press would emphasise the positive aspects of the industry, i.e. employment and career opportunities and technical innovation, and would use case studies of high-profile construction projects. A National Construction Week was launched to promote the entire industry. This would involve high-profile press conferences, site open days and events involving collaboration across the various professional bodies and trade associations.

The success of these initiatives is frustrated by the continued presence

of so many different organisations which represent this highly frag
mented industry. The lack of a clear, co-ordinated and consistent message
means that the media's focus on the negative impact of construction still
influences the opinions of the majority of people.

This book has used a combination of theory and practical real-life
illustrations to highlight key principles for the more effective management of public relations strategies for improving corporate communications at the business and project level in construction. This final chapter
will review the 'who, what and how' of corporate communications,
summarise the practical tools developed in the text, and consider the basic
lessons for construction.

Who do we need to communicate with?

Chapter 1 introduced stakeholder theory, which helps organisations
identify those groups that have an influence on the organisation and those
that they need to influence. The chapter noted that not all stakeholders are
equally important. Some have a primary influence on business decisions
and some a secondary influence. But the importance of stakeholders is not
static and may change over time.

Adopting a stakeholder approach to business ensures that companies
question their own value system to achieve a better fit with the
environment in which they operate. It means that appropriate relationship building and communication are directed at particular stakeholders.
The techniques and approaches used for each group of stakeholders will
therefore be more appropriate.

What do we need to communicate?

Construction organisations need to communicate a distinct corporate
identity and image. For a corporate identity to be more than a cosmetic
exercise and to give real support to a business and provide a competitive edge, it needs to be controlled, steered and co-ordinated. It has to
be backed up by real and positive changes in behaviour by all managers and staff. A strong corporate brand identity can differentiate the
organisation in the minds of clients and create a sense of loyalty among
the company's own staff. Corporate identity is a catalyst for change and
can encourage and promote cohesion in increasingly larger and indeed
global corporations, where there is a need to establish a sense of
belonging.

what it needs to communicate, it then has to establish clear and measurable corporate communications objectives for internal and external publics. The corporate communications plan should support more direct marketing efforts by improving relations with customers. It should ensure financial stability by communicating more effectively with investors and the financial community. The plan should also enable the organisation to relate more positively with the local communities in which it operates and serves.

The organisation needs to carefully analyse corporate communications environments, i.e. clients, employees, suppliers and subcontractors, shareholders, government, professional bodies, and the print and broadcast media. The information which the company gathers will be essential in enabling it to make choices of strategy and techniques, for example whether to manage corporate communications in house or through outside agencies, and the most appropriate tools and techniques to be used.

Those charged with the function of corporate communications will need to gain top leadership support and involvement. Feedback too will be essential. Have corporate communications efforts improved recognition and relationships with clients, investors, employees and communities?

Communication with the principal stakeholders

This book has introduced practical approaches to communicating with clients, the media, investors and financiers, government and communities.

Communicating with the client

A corporate communications programme can help to develop and sustain a company's reputation with the customers and clients, and support more direct marketing approaches. It is essential to identify who the clients are, and what they expect in terms of the quality of service and end-product. It is also essential to develop a clear understanding of the demands of the particular contract.

An effective approach to client services to beat the competition in the pre-qualification phases is important. This will require some central source of information on past and current clients and projects, and clear communications across the functions and regions of the organisation. During the construction phase of the contract it is important to ensure open channels of communication with the client to reduce conflict between the parties. An effective complaints handling procedure should encourage clients and their advisors to communicate their criticisms and views to help in improving its services. After-care services should be developed to continue to satisfy the client after the building or structure has been handed over. Customer/client care programmes generate an internal customer-orientated culture within the organisation and may be used as a marketing tool.

The three Ts, targeting, tailoring and trimming, have been identified as essential rules in the provision of promotional material to what are generally highly experienced and technical audiences. It needs to be targeted at those within the client team who will be involved in the decision-making process; to be tailored to the client's specific requirements; and to be trimmed, i.e. concise and to the point.

The public relations function should assist more direct marketing activities by keeping the organisation close to its clients, and by building long-term relationships.

Approaching the media

There is a need to establish clear and measurable objectives for any media campaign. Information and news provided to media editors and journalists should be in a user-friendly style. An effective report to the media will clearly identify who has made the news, what has happened, why it is considered to be important, and where and when the event occurred. Effective media relations will help in a crisis if appropriate relations have been developed with journalists.

Efforts should be made to evaluate the success or failure of media campaigns to ensure that objectives are being achieved.

Communicating with investors and financiers

Investor and financial relations is a specialist area. The stakeholders involved in this particular area are more demanding and critical in their approach to the company. They expect communications to be prompt and unambiguous. There is more regulation of financial communications particularly for listed companies. The use of the annual report as a communication tool should be carefully considered in terms of design and content.

...... is a need for the organisation to understand how decisions are reached throughout the governing process. Parliamentary interaction is not sufficient: the executive also needs approaching.

Research and preparation are fundamental to interacting with government. Use of up-to-date and accurate information is essential to be successful. Building relations and establishing proper credentials is a long, time-consuming process.

Communicating with communities means business

The communities in which construction companies operate are critical for business and project success. Construction community involvement can be approached at two levels, corporate and project level, and should be based on a focused approach which has a strong business case.

Companies need to act responsibly, and this needs to be communicated. Every action that is taken should be designed to have a positive impact on the local community and improve the quality of life for everyone. Clients increasingly expect organisations to adopt a community approach to projects but companies need to ensure that the client has attended to all community issues before they start work on a contract. Construction organisations need to interact closely with the client from day one, and to be aware of all community issues: the client will be aware of all issues that have been raised during the planning process. Construction companies should be proactive and appoint project liaison engineers and organise forums with local people and other interested parties.

Communicating with our own people

The successful internal communications of construction organisations are critical to the effectiveness and success of the management of that company, and therefore to the success of that business itself. Communication of key corporate messages concerning safety, quality or environmental aspects is fundamental to how staff work, how they perform and how much they commit themselves to the common cause. A company's very culture is dependent upon its internal communication.

The main objective is to motivate and inform company staff in order to help achieve the overall business goals successfully. Internal communications must meet the corporate objective, i.e. that of giving out serious management messages for future corporate development. This could be through a variety of media and concern any aspect of the corporate strategy.

The communications role of the project manager

The manager of the construction site has an important role to fulfil in the overall business success of the company. Involvement starts from the bidding process and continues throughout the life of the project.

As a communicator the project manager has a far-reaching influence on the way the project is portrayed. Planning is a critical function for site communication. There is a need to establish a communication plan with a contingency view. Establishing a project brand will assist in establishing a more cohesive and effective construction team.

Increasingly, the project management team is engaged in communicating with prospective clients during the pre-qualification process. The team needs to pitch the message to the correct audience. This means understanding who they are communicating with and their backgrounds, needs and interests. Everyone needs to be 'on message' during interviews, presentations, events, visits, etc. Preparation is crucial. Positive messages should be generated about the project which can be used in wider marketing communications efforts.

Safety communication with our own staff and the community is vital

Safety is a major element that contributes to the reputation of a construction organisation. There needs to be a practical partnership between the safety department and public relations. The aim has to be to comply with the legal obligations and to create, protect and enhance a safety culture. Communications efforts include targeting managers and staff on site, the press and media and local communities. Public relations needs to get involved with the induction of new recruits and training programmes. A good safety record and a comprehensive approach to safety training are important marketing tools which need to be communicated to existing and potential clients.

Managing incidents so that mole hills don't become mountains!

All companies have to deal with incidents of varying degrees of severity from time to time. From a change of chairman or CEO, to an aggressive take-over bid or an accident on site, incident management is about being in control and having established guidelines and mechanisms for dealing with incidents of whatever nature. All parties need to be informed in advance about such procedures and to know what role and responsi-

way in which important stakeholders view the organisation.

Practical communication tools

Some practical communications tools and techniques identified in this book include:

- Stakeholder identification and analysis to ensure more effective targeting of communication.
- Implementation of corporate identity programmes as catalysts for positive change in construction organisations.
- Development of communication and public relations skills and formation of dedicated functions within construction organisations.
- Development of client care programmes and more effective marketing communications techniques.
- Approaches to the creation of improved media relations.
- The design of more effective annual reports and accounts and the building of better relationships with the financial community.
- Techniques aimed at making the case for construction with government.
- Development of community policies which can improve the image of the construction business, win work and help develop the skills of the construction team.
- Developing the public relations skills of the project management team, especially in relation to dealing with clients and in managing site events.
- Practical communication of safety policies and programmes inside and outside the construction company which can help avoid accidents and the consequent damaging publicity which can affect marketing success.
- Incident management plans which ensure that the construction organisation is in control, and communicates an image of being in control!

Some basic lessons learnt

Construction organisations need to:

- **Say *and* do the right thing!** The image that they communicate needs to reflect reality. It needs to build on the corporate culture of the organisation and provide a distinct identity with clients, managers and staff and the communities in which it operates

- **Do their homework and plan, plan, plan!** They need to find out as much as possible about the people and organisations that they need to communicate with, and to develop appropriate relationship-building techniques. They need to organise themselves and plan their approach to the communications aspects of business and project management.
- **Develop everybody's communications skills!** Every manager and member of the team needs to understand that everything they do and say affects the corporate communication and image being projected. All construction people need to be trained in how to communicate more effectively with each other, and with the outside world. It could mean the difference between winning or losing a contract.

This book has not attempted to solve all the industry's problems, but has addressed the need for organisations in construction to manage better the way in which they communicate with the environments in which they operate. The cumulative effect of businesses taking a more strategic view of communications at the corporate and project level should contribute to an improvement in the perceptions that clients, the media, government, financiers and communities have of this very important industry.

References

Chapter 1

Ackoff, R. (1974) *Redesigning the Future*. John Wiley, New York.

Ansoff, I. (1965) *Corporate Strategy*. McGraw-Hill, New York.

Freeman, R.E. (1985) In *Strategic Management: A Stakeholder Approach* (Marschfield, M.A., ed.). Pitman, London.

Post, J. (1981) Research in business and society: current issues and approaches. *AACB Conference on Business Environment and Public Policy*, Berkeley, CA.

Smith, C. (1994) The new corporate philanthropy. *Harvard Business Review* May–June, pp. 105–16.

SRI (1963) Stakeholder management: A case study of the US Brewers and the Container Issue. *Applications of Management Science* **1**, 57–90.

Chapter 2

New Civil Engineer (1996) 2 May.

PR Week (1997) 20 June.

Chapter 3

Sheldon Green, P. (1994) *Winning PR Tactics: Effective Techniques to Boost Your Sales* pp. 142–45. Pitman Publishing, London.

White, J. (1991) *How to Understand and Manage PR*. Business Books, p. 146.

Chapter 4

American Marketing Association (1960) *Marketing Definitions: A Glossary of Marketing Terms; Committee on Definitions*. AMA, Chicago.

Baron, S. & Harris, K. (1995) *Service Marketing, Text and Cases*. Macmillan, London.

Barsky, J.D. (1995) *World-Class Customer Satisfaction*. Irwin Professional Publishing, New York.

Bly, R.W. (1995) *Create the Perfect Sales Piece, A Do It Yourself Guide to Promotional Brochures, Catalogues, Fliers and Pamphlets*. John Wiley, New York.

Burnett, J.J. (1988) *Promotional Management: a Strategic Approach*, 2nd edn. West Publishing Company, St Paul.

Clutterbuck, D. (1988) Developing customer care training programmes. *Industrial and Commercial Training* **20** (1), 11–14.

Clutterbuck, D. & Kernaghan, S. (1991) *Making Customers Count: a Guide to Excellence in Customer Care*. Mercury Books, London.

Contract Journal (1998) Launch to boost contractor's profile. 14 January.

Cook, S. (1992) *Customer Care: Implementing Total Quality in Today's Service-driven Organisation*, Kogan Page, London.

Davis (1988) *Business-to-business Marketing and Promotion*. Business Books.

Dibb, S., Simkin, L., Pride, W.M. & Ferrell, O.C. (1994) *Marketing: Concepts and Strategies*, 2nd ed. Houghton Mifflin, London.

Fellows, R. & Langford, D. (1993) *Marketing and the Construction Client*. Chartered Institute of Building/Bourne Press, Bournemouth.

Fisher, N. (1986) *Marketing in the Construction Industry: A Practical Handbook for Consultants, Contractors and Other Professionals*. Longman Group, Harlow.

Ind, N. (1990) *The Corporate Image: Strategies for Effective Identity Programmes*. Kogan Page, London.

Jefkins, F. (1996) *Public Relations*, 2nd edn. Butterworth Heinemann, Oxford.

Kotler, P. & Armstrong, G. (1993) *Marketing: An Introduction*, 3rd edn. Prentice-Hall, Englewood Cliffs, NJ.

Morgan, R.E. & Morgan, N.A. (1991) An exploratory study of market orientation in the UK consulting engineering profession. *International Journal of Advertising* **10**, 333–47.

Nickels, W.G. (1984) *Marketing Communication and Promotion*, 3rd edn. Grid Publications, Columbus.

Olins, W. (1978) *The Corporate Personality: an Enquiry into the Nature of Corporate Identity*. The Design Council, London.

Olins, W. (1989) *Corporate Identity: Making Business Strategy Visible through Design*. Thames & Hudson, London.

Olins, W. (1990) *The Wolff Olins Guide to Corporate Identity*. The Design Council, London.

Pearce, P. (1992) *Construction Marketing: a Professional Approach*. Thomas Telford, London.

Pollock Nisbet Partnership (1994) *Customer Care and Housing Management: a Research Report*. Scottish Homes, Edinburgh.

Preece, C.N. & Moodley, K. (1995) The management of direct selling and pre-qualification team presentations for competitive advantage in contractual services. *Conference on Practice Management for Land, Construction and Property Professionals*. Liverpool, Sir John Moores University, 5 October.

Preece, C.N., Moodley, K. & Habeeb, M. (1995) The management of direct selling and pre-qualification team presentations for competitive advantage in con-

Hill, New York.

Sheldon Green, P. (1994) *Winning PR Tactics: Effective Techniques to Boost Your Sales.* Pitman Publishing, London.

Smith, A.M. & Lewis, B.R. (1989) Customer care in financial service organisations. *International Journal of Bank Marketing*, **7** (5), 13–22.

Storback, K., Straandvik, T. & Gronroos, C. (1995) Managing customer relationships for profit: the dynamic of relationship quality. *International Journal of Service Industry Management.*

Thompson, N.J. (1996) Relationship marketing and advocacy. In *1st National Construction Marketing Conference Proceedings*, 4 July, Oxford Brookes University, pp. 78–90.

Woodruffe, H. (1995) *Services Marketing*. Pitman, London.

Chapter 5

Black, S. (1995) *The Practice of Public Relations*, p. 13. Butterworth-Heinemann, London.

Dibb, S., Simkin, L., Pride, W.M. & Ferrell, O.C. (1994) *Marketing: Concepts and Strategies*, 2nd edn, p. 429. Houghton Mifflin, London.

Institute of Public Relations (1995) *Professionalism in Practice*, p. 4.

Ridgeway, J. (1996) *Practical Media Relations*, p. 5. Gower, London.

Smith, P.R. (1994) *Marketing Communications – An Integrated Approach*, p. 276. Clays.

Chapter 6

Greenbury Committee (1995) *Directors' Remuneration*. Gee Publishing.

Jefkins, F. (1994) *Public Relations Techniques*, 2nd edn. Butterworth-Heinemann, London.

Chapter 7

Miller, C. (1990) *Lobbying Government: Understanding and Influencing the Corridors of Power*, 2nd edn. Blackwell, Oxford.

Chapter 9

Adams, R., Curruthers, J. & Hamill, S. (1991) *Changing Corporate Values. A Guide to Social and Environmental Policy and Practice in Britain's Top Companies.* Kogan Page, London.

Avashi, B. (1994) What is business's social compact? *Harvard Business Review* January–February, 38–48.

Costain Group Publications (1994a) *Community Involvement Policy.*

Costain Group Publications (1994b) *Costain and Young People Joint Venture.*

John Laing Group plc (1986–1996). Annual Reports.

Kanter, R.M. (1995) *World Class – Thriving Locally in a Global Economy*, pp. 29–33, 174–97. Simon and Schuster, New York.

Moodley, K. & Preece, C.N. (1996) Implementing community policies in construction. In *The Organisation and Management of Construction. Shaping Theory and Practice* (Langford, D. & Retik, A., eds). E & F N Spon, London.

Smith, C. (1994) The new corporate philanthropy. *Harvard Business Review* May–June, 105–16.

Chapter 10

American Project Management Institute Standards Committee (1996) *A Guide to the Project Management Body of Knowledge.* American Project Management Institute.

Association of Project Managers (1996) *The APM Body of Knowledge.* Association of Project Managers.

Baker, B. & Murphy, D. (1988) Factors affecting project success. In *The Project Management Handbook*, 2nd edn (Cleland, D.I. & King, W.R., eds). Van Nostrand Reinhold, New York.

Kerzner, H. (1992) *Project Management: a Systems Approach to Planning, Scheduling and Controlling*, 4th edn. Van Nostrand Reinhold, New York.

Turner, J.R. (1995) *The Commercial Project Manager: Managing Owners, Sponsors, Partners, Supporters, Stakeholders, Contractors and Consultants.* McGraw-Hill, New York.

Wilerman, D. & Baker, B. (1988) Some major research findings regarding the human element in project management. In *The Project Management Handbook*, 2nd edn (Cleland, D.I. & King, W.R., eds). Van Nostrand Reinhold, New York.

Chapter 12

PR Week (1997a) 4 April, Haymarket Business Publications, London.

PR Week (1997b) October, Haymarket Business Publications, London.

Construction Industry Board (1996) *Constructing a Better Image*. A Report by Working Group 7 of the Construction Industry Board. Thomas Telford.

Latham, M. (1994) *Constructing the Team*. HMSO, London.

Bibliography

Abratt, R. (1989) A new approach to the corporate image management process. *Journal of Marketing Management* **5** (1), 63–76.

Bell, R. (1981) *Marketing and the larger construction firm*. Occasional paper number 22. Chartered Institute of Building, London.

Bernink, B. (1995) Winning contracts. In *The Commercial Project Manager* (Turner, J.R., ed.), pp. 325–49. McGraw-Hill, New York.

Bland, M. & Mondesir, S. (1987) *Promoting Yourself on Television and Radio*, p. 7. Kogan Page, London.

Burnett, J.J. (1988) *Promotion Management; a Strategic Approach*. West Publishing Company, New York.

Cadbury Committee (1995) *The Financial Aspects of Corporate Governance – Compliance with the Code of Best Practice*. Gee Publishing.

Cooper, P.D. & Jackson, R.W. (1988) Applying a services marketing orientation to the industrial services sector. *Journal of Services Marketing* **2** (4), Fall, 67.

Courtis, J. (1987) *Marketing Services: a Practical Guide*. British Institute of Management/Kogan Page, London.

Davis, S. (1988) *Business-to-business Marketing and Promotion*. Business Books.

Dubs, A. (1988) *Lobbying*. Pluto Press, London.

Gerwick, B.C. & Wooley, J.C. (1983) *Construction and Engineering Marketing for Major Project Services*. John Wiley, New York.

Gray, E.R. & Smeltzer, L.R. (1985) SMR forum: corporate image – an integral part of strategy. *Sloan Management Review* Summer, 73–78.

Haywood, R. (1984) *All about PR*. McGraw-Hill, Maidenhead.

Ind, N. (1990) *The Corporate Image: Strategies for Effective Identity Programmes*. Kogan Page, London.

Jefkins, F. (1990) *Modern Marketing Communications*. Blackie, London.

London Stock Exchange *A Guide to Going Public*.

London Stock Exchange *Guidance on the Dissemination of Price Sensitive Information*.

Longhurst, J. (1991) Report design: handle with care. *Accounting* April, 102–4.

Moore, A.B. (1984) *Marketing Management in Construction; a Guide for Contractors*. Butterworth, London.

Nickels, W.G. (1984) *Marketing Communication and Promotion*, 3rd edn. Grid Publications, Columbus.

Olins, W. (1990) *The Wolff Olins Guide to Corporate Identity*. Design Council, London.

..., C.N. (1996) Keeping up appearances; the planned and pro-active management of public relations for contractors involved on transport projects. *Motorway and Trunk Road Contracting, Finance & Maintenance. SMI Conference,* London, 10–11 March.

Preece, C.N. & Male, S.P. (1997) Promotional literature for competitive advantage in UK construction firms. In *Construction Management and Economics* (Bon, R. & Hughes, W., eds). E & F N Spon, London.

Preece, C.N. & Moodley, K. (1995) Cost effective promotional strategies for small firms in developing countries. *CASLE/FIG Seminar on Sustainable Development: Counting the Cost and Maximising the Value,* Harare, Zimbabwe, 13–17 August.

Preece, C.N. & Moodley, K. (1995) Escape from the image trap. *Chartered Builder* September, 11–13.

Preece, C.N. & Moodley, K. (1996) Developing a programme of total customer satisfaction through service quality for competitive advantage in construction. *CIB W55 Economic Management and Innovation, Productivity and Quality in Construction,* Zagreb, Croatia.

Preece, C.N. & Moodley, K. (1996) The management of public relations as a business process in construction. *CIB Working Commission 65 Organisation and Management of Construction (in Collaboration with W 92) Symposium Shaping Theory and Practice.* Glasgow, 28 August–3 September.

Preece, C.N. & Moodley, K. (1996) Winning pitches. *Construction Manager* **2** (2), 16–19.

Preece, C.N. & Moodley, K. (1996) The management of public relations as a business process in construction. In *The Organisation and Management of Construction; Shaping Theory and Practice, Vol. 1 Managing the Construction Enterprise* (Langford, D.A. & Retik, A., eds), pp. 303–11. E & F N Spon, London.

Preece, C.N. & Tarawneh, S.A. (1996) Re-orientating the construction team to achieve service quality for client satisfaction. *First National Construction Marketing Conference,* Oxford Brookes University, 4 July.

Preece, C.N. & Tarawneh, S. (1997) Developing high quality design and build services: closing the four managerial gaps. *Second National Construction Marketing Conference,* Oxford Brookes University, 3 July.

Preece, C.N. & Tarawneh, S. (1997) Service quality for client satisfaction on design and build projects. *Association of Researchers in Construction Management (ARCOM) 13th Annual Conference Proceedings,* King's College, Cambridge, 15–17 September.

Preece, C.N. & Tarawneh, S. (1997) Why are design and build clients unhappy? *Construction Manager* **3** (7), 24–25.

Preece, C.N., Bissonauth, M. & Moodley, K. (1996) The application and effectiveness of quality assurance in small contracting firms – an empirical study. *CIB W55 Economic Management and Innovation, Productivity and Quality in Construction.* Zagreb, Croatia.

Preece, C.N., Bissonauth, M. & Moodley, K. (1997) The management of quality in small contracting firms – an empirical study. *Fourth International Conference on Civil Engineering,* 4–6 May. Sharfi University of Technology, Tehran.

Preece, C.N., Male, S. & Moodley, K. (1996) Targeting, tailoring and trimming – the more effectiveness of promotional literature in the marketing of contractual services. *First National Construction Marketing Conference*, Oxford Brookes University, 4 July.

Preece, C.N., Moodley, K. & Graham, C. (1996) Public relations to support more effective marketing in UK construction contractors. *First National Construction Marketing Conference*, Oxford Brookes University, 4 July.

Preece, C.N., Moodley, K. & Graham, C. (1996) The public relations role of the project manager on transport infrastructure schemes. *24th PTRC European Transport Forum*, 2–6 September.

Preece, C.N., Moodley, K. & Habeeb, M. (1996) The management of direct selling and prequalification team presentations for competitive advantage in construction. In *Practice Management for Land, Construction and Property Professionals* (Greenhalgh, B. (ed.), pp. 129–37. E & F N Spon, London.

Preece, C.N., Moodley, K., Humphrey, J. (1997) Effective media relations strategies for civil engineering and building contractors. *Second National Construction Marketing Conference*, Oxford Brookes University, 3 July.

Preece, C.N., Moodley, K. & Humphrey, J. (1997) Effective media relations strategies for contractors involved with transport projects. *25th PTRC European Transport Forum*, Brunel University, 1–5 September.

Preece, C.N., Moodley, K. & Putsman, T. (1997) More effective pre-qualification strategies and team presentation in contractual services. *CIB W92 Conference Procurement – A Key To Innovation*, Montreal.

Preece, C.N., Putsman, A. & Walker, K. (1996) Satisfying the client through a more effective marketing approach in contracting. *Oxford Brookes University 1st National Construction Marketing Conference Proceedings*, 4 July, pp. 5–9.

Preece, C.N., Shafiei, M.W.M. & Moodley, K., (1997) The effectiveness of promotion in the private house building industry. *Second National Construction Marketing Conference*, Oxford Brookes University, 3 July.

Preece, C.N., Tarawneh, S. & Moodley, K. (1997) Modelling design and build service quality. *International Conference on Leadership and Total Quality Management in Construction and Building*. Institution of Engineers, Singapore, 6–8 October.

Rushton, A.M. & Carson, D.J. (1985) The marketing of services: managing the intangibles. *European Journal of Marketing* **19** (3), 19–40.

Stanley, R.E. (1983) *Promotion: Advertising, Publicity, Personal Selling, Sales Promotion*. Prentice-Hall, Englewood Cliffs, NJ.

Stewart, K. (1991) Corporate identity: a strategic marketing issue. *International Journal of Bank Marketing* **9** (1), 32–39.

Topalian, A. (1984) Corporate identity: beyond the visual overstatements. *International Journal of Advertising* **3** (3), 55–62.

Index

accidents 161, 170–71, 173–6, 200
accounts 50–51, 76–7
achievement goals 18–19, 38–9
advertising 68
advocacy, government relations 85–6
annual reports 50–51, 76–7
articles, feature 66–7
audiences, identification 62, 63

banks 75
benefits, client care 48
booklets 161–2
brand identity 24–6
briefings 24, 182–3
brochures, marketing 51–2
business cards 23
business cases 105–7

caring, for clients 43–59, 103
CDM, *see* Construction (Design and Management) Regulations
challenges, in/outside organisation 34
change 9, 19–20
children 176–7, 188–90
    see also schools
CIMAH, *see* Control of Industrial Major Hazards
clients 55, 207–8
    care 43–59, 103
    needs 46–7
The Commission (EC) 96–9
communications
    definition xi
    techniques 49–51, 65–8
community
    business cases 105–7
    gaining support 108–9
    industry relations 102–3

involvement 100–116
operational issues 109
policies 104–5, 107, 110–15
relations 100–116
resource commitment 108
service orientation 113–14
company manual 126
competition 17–18, 45–7, 53, 106
computers 29
    see also personal computers
conferences 50–51, 67–8
Considerate Constructors Scheme 205
Construction (Design and Management) Regulations (CDM) 156–7
Construction (Health, Safety and Welfare) Regulations (1996) 156, 164–5
Construction Industry Board Working Group 7 205
contact, impersonal 90–91
contingency management 144
contractual services 53
Control of Industrial Major Hazards (CIMAH) 195
corporate culture 120–21
corporate identity 16–32, 55–6, 206
    house style 23–4
    importance 25–6
    making it work 28–30
    schemes 24–6, 50
    strategic tool 16–17
Council of Ministers 95
crises
    definition 192–3
    management 173–6, 192–201
        see also incident management
    media assistance 68–9

customer care 25, 49–50
  see also clients, care
customer services 113–14

direct action 114–15
dissatisfaction, clients 46

E-mail, *see* electronic mail
economic influence 9
education 110, 112–13
  see also schools
effectiveness, monitoring 80–1
electronic mail (E-mail) 134
electronic media 65
  see also World Wide Web
emergencies, *see* incident management
employees 75, 108–9, 119–35
employers in the commnunity 112
enterprise 112–13
environment 104–5, 110–11
ethical investment 107
Europe 94–9
event management 144
executive functions 36
Exxon-Valdez case 102

face-to-face communications 131–3
faxes 30–31, 135
feature articles 66–7
finance 72–8, 208
flow charts 197–8
function of communications 36–7

goals 18–19, 38–9, 121–3
  see also objectives
government 79–99, 209
  common mistakes 93–4
  contacting targets 87–91
  decision takers 80
  information sources 81–5
  internet addresses 83–4
  making your case 80–81
  meetings 91–2
  officials 88–90
  outside help 94
  publications 82–3
  visits 92–3
green issues 12
  see also environment

handbooks 161–2

hazards 162–0
HBG Construction Limited 26–8
Health and Safety at Work Act 1974
  155–6
Health and Safety Executive (HSE) 154,
  155, 162
house journals 50
house style 16, 22, 23–4
HSE, *see* Health and Safety Executive

identity, *see* corporate identity
IIP, *see* Investors in People
image 21–2, 105–6
incidents
  communication plan 195, 197–8
  definition 192–3
  flow chart 197–8
  management 173–6, 210–11
  planning for 195–201
  reaction to 198–9
induction 123–6, 161
industry 70–71, 104–5
  policies 107, 110–15
information
  management 144
  online 56–8, 65, 83–4
  sources 81–5
innovation 25
institutions 74
insurance, PR crises 193–4
internal communications 119–35
  corporate culture 120–21
  goals 121–3
  induction packages 123–6
  techniques 123–31
  verbal presentations 131–5
internal management 145, 159
internet, *see* World Wide Web
interviews 56, 63, 146
investments 74, 107
investors 77, 208
Investors in People (IIP) scheme 122,
  162

journalists 174, 180–81

law, *see* legislation
leaflets 161–2
legislation 93, 155–7
  see also Construction Regulations,
  Health and Safety

line managers, induction duties 125
listed companies 75–6
literature
    marketing 51–2
    promotions 53–6
lobbying 114–15
local community 183–8
logos 16–32

magazines 64
making your case, government
        relations 80–81
management
    communications 33–9
    conferences 50–51
    functions 36
    incidents 210–11
    philosophies 10–11
    relationships xi–xii
    skills 35
    stakeholders 4–5
    values 13–14
mapping, stakeholders 11
marketing
    see also public relations
    brochures 51–2
    communications 37, 43–59
    competition 53
    definition 43–4
    external communications 158
    partnering 44–5
    public relations 43–4
    relationships 44–5
    techniques 50–51
media 159, 208
    see also press
    early involvement 179
    outlets 63–8
    relations 60–71
        crises 68–9
        definition 61–2
        evaluation 69
        strategies 62–3
    reporting incidents 174
    safety issues 171–83
Members of Parliament (MPs) 88–90, 95
mergers 26–8
money, see finance
monitoring, effectiveness 80–81
MPs, see Members of Parliament
Murphy's law 142

New Corporate Philanthropy 101–2
newsletters, safety 158, 163–7
newspapers 60–71
    staff 126–30, 158, 159, 168–71
noticeboards, sites 134

objectives
    communications 33–9, 121–3
    corporate 122
    identification 62
online information 56–8, 65, 83–4
open evenings 186–8
opportunities, analysis 107–8
organisation, communications 33–9
outside help 38–9

Parliaments 88–90, 95, 97–8
partnerships 44–5, 153
PCs, see personal computers
personal computers (PCs) 29, 124, 134
personal contact 88–90
photographs 66, 180–81, 182
police force, safety talks 189–90
policies
    community 104–5, 110–15
    environment 104–5, 110–11
    regeneration 111
political agenda 10
posters, safety 158, 168
power 79–99
pre-qualification process 51–2, 56,
        146–8
presentations 56, 131–5, 146
press 65–8, 74, 159
    local 176–9
    releases 177–8
    reporting incidents 174
    technical 64, 180–83
projects 139–51
    corporate-level support 145
    cycles 142–3
    execution 148–9
    interaction 142–3
    management 210
    public relations 143–4
    stakeholders 139–40
    strategies 105, 140
    teams 146–8
promotions 50, 53–6, 93
public forums 186–8
public opinion 9

public relations
  communications 35–6
  communications purpose 119
  corporate identity 20–21
  crises 198, 199
  function, media outlets 172
  manager obligations 172
  marketing 43–4
  projects strategy 143–4
  safety 153–5, 157–9
  stakeholders 3–4
publications 82–3, 130–31

quality 19, 25, 45–7, 49–50
questions flow chart 197–8

radio 64–5, 67
recruitment 106–7
regeneration policies 110, 111
relationships
  building 11–12
  community 100–116
  finance 72–8
  government 79–99
  investors 77
  management xi–xii
  marketing 44–5
  newspapers 60–71
reporting 174, 180–81
  see also press and media
resources, commitment 108
response documents 175

safety 152–91, 210
  competitions 163
  induction talks 161
  inspections 163
  law 155–7
  newsletters 158
  posters 158
  press and media 171–83
  programmes 171
  public relations 157–9
  talks in schools 176–7, 188–9
school liaison 159, 176–7, 188–9
sensationalisation 180–82, 193, 194–5
service 45–7, 49–50, 113–14

shareholders 73–4
sites, safety 149–50, 196
skills
  communications 33–9
  project management 141–2
social clubs 134
sponsorship 68
stakeholders 3–15
  communication 207–11
  financial 73–5
  identity 7–8
  management 4–5, 6–7
  mapping 11
  primary 5–6
  production view 4
  projects 4–5, 139–40
  public relations 3–4
  secondary 6
  the stake 8–11
  values 12–13
statements 198–9, 200
strategies
  corporate identity 16–17
  different client types 55
  media relations 62–3, 69
  projects 105, 140
style, corporate identity 55–6

takeovers 26–8
targeting 54, 62, 85–6
Tarmac World 128–9
technology 9, 20–24, 29
television 64–5, 67
third-party questions flow chart 197–8
tools, communication 211
training 110, 112–13, 162
turf-cutting ceremonies 184–6

value analysis 14
values 12–14
visits, government relations 92–3

warning signs, hazards 176
web sites 56–8
word of mouth 55
World Wide Web 56–8, 65, 83–4